PRACTICAL
PROGRAMMING

S.M.P. HANDBOOKS

We Built our own Computers

PRACTICAL PROGRAMMING

BY P.N.CORLETT AND J.D.TINSLEY

CAMBRIDGE
AT THE UNIVERSITY PRESS
1968

Published by the Syndics of the Cambridge University Press
Bentley House: 200 Euston Road, London, N.W.1
American Branch: 32 East 57th Street, New York, N.Y. 10022

Library of Congress Catalogue Card Number: 68–21391
Standard Book Numbers:
521 07261 1   clothbound
521 09542 5   paperback

Printed in Great Britain by
Alden & Mowbray Ltd
at the Alden Press, Oxford

FOREWORD

I used to think that the teaching of numerical analysis in a university mathematical syllabus could be done by the sole use of 'epsilons' and that the subject is as easy as classical analysis to assimilate in this way. Experiment has proved me wrong, and there is little doubt that most students need some extra stimulus before they can get excited about practical mathematics. After the first undergraduate year it is almost too late to stimulate this excitement, because pure mathematics has captured the imagination and there is no real motivation for doing anything else.

This book is therefore particularly important for me, because the knowledge of what a computing machine can do, and how one gets it to do it, will have been transmitted to the student reader *before* he comes to the University. He will then already have some motivation for further study, and at the Freshman stage will appreciate the fact (which teachers should make clear at school) that the study of more advanced mathematics is a necessary preliminary to the subsequent application of his knowledge of machines and programming in a really expert way and to really important problems.

Schools, of course, are concerned not solely, nor even essentially, with students who will become professional mathematicians with future university training. Many school-leavers seek employment directly, and for them a knowledge of computing is a passport to an interesting and indeed lucrative career. Others may be scientists, who use mathematics as a tool. They, above all, will have to use machines to solve their ultimate problems, and will find the material of this book of significant value, as well as interest, even in their early training. Some students may not be particularly interested in mathematics or science, but the computer is a fact of modern life, and here, at an age when everything is important, they can learn about an invention which will certainly affect their future. Finally, in a period in which we hear much about the dangers of too-early specialization, and when 'embellishment' is a U word, we have in this book something which delays the former and which achieves the latter, perhaps particularly in mathematics.

The material is not just a factual account of Algol programming and worked examples in arithmetic. The reader is constantly urged to think; the worked examples cover a wide range of both numerical and non-numerical problems (though the former are bound to predominate at this stage and with the Algol language), and the exercises both consolidate and extend the reader's knowledge.

I therefore warmly welcome this addition to the S.M.P. series of books, and am very glad to have been able to assist the authors, in a small way, by making available to them and their students, and for that matter those of other local schools, the facilities at the Oxford Computing Laboratory.

L. FOX, D.Sc.
Professor of Numerical Analysis
Oxford University

CONTENTS

CONTENTS

Programs are normally printed in flow diagram form
and accompanied by a version in KDF 9 Algol

† denotes that the flow diagram is omitted

* denotes that an Algol version is not printed

PREFACE

This book was written for those wishing to make use of the many computing facilities now available to students. It is hoped that it will ease the way for colleges and schools starting computing in the near future, and will act as a stimulus for those who had not thought such work within their capabilities. It may also appeal to those requiring an insight into practical computing methods before embarking on their own research.

Because of its importance in scientific and mathematical work, the language used throughout the book is Algol. The majority of the programs are also given in flow diagram form so that they may be written in other computer languages. The text sets out both to explain the necessary mathematical background and to provide a commentary on the chosen examples. It is hoped that the completion of these will provide at least a glimpse into the real world of programming, and that an appreciation may be gained of the power and the limitations of the computer.

The inspiration for this book comes from an attempt to teach the elements of computer programming to boys of St. Edward's School, and the project has been greatly assisted by Professor Fox and the staff of the Oxford University Computing Laboratory, where the examples were validated on an English Electric KDF 9. The authors wish to acknowledge the help of A. J. Cox, J. M. F. Davidson, M. I. Caffyn and others in the Sixth Form at St. Edward's who have offered original ideas and written many of the programs. We would also like to thank Dr. C. E. Phelps and Mr. H. D. Williams for reading the proofs and offering valuable suggestions, and Dr. D. C. Handscomb for permission to use his procedure Randbody in program 7.3. Finally, we would like to thank Mrs. J. D. Tinsley for her painstaking work with the manuscript, the diagrams and with the proof reading.

P. N. CORLETT   J. D. TINSLEY

1

INTRODUCTION

The computer is a very new member of the family of calculating devices, but its development has been extremely rapid. The abacus, logarithmic tables, slide rule and desk calculating machine all depend on the human operator for their manipulation. The computer, however, is a calculating device which can be made to follow a sequence of instructions automatically, and at great speed. This sequence of instructions is called a program (using the American spelling), and once the program has been prepared, it may be used either for a single calculation or for a series of similar calculations on different data.

Programs for early computers had to be written in a machine code in which every instruction was made up from a series of binary digits, 0 and 1. The preparation of machine-coded programs is a long and tedious process, and methods have been developed to reduce the labour involved. Programs for a modern computer are written in languages more closely resembling the ordinary language of mathematics and are easy to understand and simple to construct. The task of translating these languages into machine code is now performed by the computer itself, using a built-in routine of instructions called a compiler.

The principal languages in use at the time of writing are Algol and Fortran, both of which have wide use in the scientific field. Languages have also been developed for special purposes; for example, Cobol, which is used in business. Earlier computers used languages called Autocodes, which were designed for a particular computer: for example, the Mercury and Pegasus Autocodes. The basic principles of computing are common to all languages, and may be described by flow diagrams which define the sequence of instructions without using a particular language.

COMPONENTS OF A COMPUTER

Computers vary considerably in design, but the basic components may be illustrated by means of the following diagram:

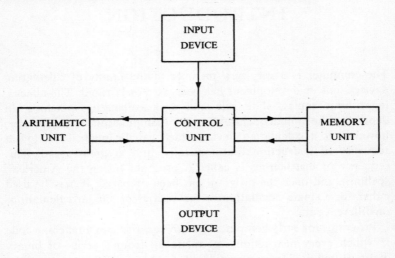

The program and data are usually prepared on punched cards or punched paper tape, using a typewriter with keyboard characters corresponding to those of a particular programming language. The most common output device is a line-printer, which prints the results of a program directly in a readable form. The memory unit is generally in two parts: an immediate access part which can be read at any time, and a backing store with a much larger capacity, but with a longer delay time for reading information. The arithmetic unit performs the basic operations of addition, subtraction, multiplication and division, and the control unit supervises the operation of the computer and follows the instructions prepared by the programmer.

NUMBER STORAGE AND
WORKING ACCURACY

Numbers can be stored in the memory cells of a computer either as integers or as real numbers. Real numbers are expressed in floating point form and this allows each number to be stored to the maximum

2

working accuracy of the computer. For example, if the computer is able to work to five significant decimal digits, the numbers

$$31.5 \quad 4163 \quad 0.0026458 \quad 34{,}789{,}345$$

will be stored as

$$0.31500 \times 10^{+2} \quad 0.41630 \times 10^{+4}$$
$$0.26458 \times 10^{-2} \quad 0.34789 \times 10^{+8}$$

Usually, however, these numbers are printed by the output device in the form

$$3.1500 \,_{10}+1 \quad 4.1630 \,_{10}+3 \quad 2.6458 \,_{10}-3 \quad 3.4789 \,_{10}+7$$

(see page 27).

In all calculations involving numbers rounded to the maximum working accuracy of the computer, there will be an inevitable loss of accuracy because of the rounding process. For example, consider a simple calculating device in which every calculation is rounded to two significant decimal digits. If the cosine rule is used to calculate the angle C of a triangle, given the lengths of the sides a = 36, b = 42 and c = 54, the calculation would proceed as follows:

$$\cos C = \frac{a^2 + b^2 - c^2}{2ab}$$

$$= \frac{(36)^2 + (42)^2 - (54)^2}{2\,(36)\,(42)}$$

Working to two sig. figs.	*Exact working*
$= \dfrac{1300 + 1800 - 2900}{3000}$	$= \dfrac{1296 + 1764 - 2916}{3024}$
$= \dfrac{200}{3000}$	$= \dfrac{144}{3024}$
$= 0.067$	$= 0.048$ (rounded to two sig. figs.)

The final value of 0.067 compares with a value 0.048 obtained by exact working in the intermediate steps.

3

Errors also occur when two nearly equal numbers are subtracted. If the working accuracy of a computer is eight significant decimal digits, the result of the subtraction

$$7389.2476 - 7389.2424$$

is given as

$$0.0052000000$$

The last six significant figures are meaningless and the answer is accurate to two significant figures only.

A further error may be introduced by the fact that the data for a problem cannot always be stored exactly, for example numbers like π, sin 25° or even 0.6 which is stored as a sequence of recurring binary digits. The problem actually solved by the computer is then slightly different from the problem given. Also, the computed solution will be different from the exact solution, and if this difference is large, the problem is said to have inherent instability (or to be ill-conditioned).

The errors arising from the arithmetical operations of the computer depend on the method used to solve the problem. Some methods are bad in this respect, and are said to suffer from induced instability (that is, they give poor results even for well-conditioned problems). Such methods should be avoided. The study of good methods for solving numerical problems forms part of the subject of numerical analysis and is virtually independent of the techniques of programming.

FLOW DIAGRAMS

The conventions used in flow diagrams are given below.

1. Assignment statements. If a value of 3 is to be stored in a memory cell labelled p, the instruction is written as

$$p := 3$$

This is read as 'p is assigned the value of 3'. If twice the value stored in cell p is then to be stored in cell q, this is written as

$$q := 2p$$

Both the above expressions are called assignment statements and, in the flow diagrams, they are set out in rectangular blocks:

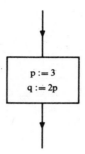

Any values previously stored in cells p and q are replaced by the values specified by the assignment statements.

2. Decision blocks. Questions are placed in blocks with curved ends. Decision blocks contain two possible exits, labelled 'yes' and 'no'.

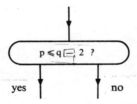

3. Input and output. Instructions to read two values from a data tape and to assign them to x and y are written as

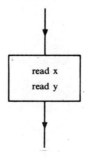

The first number on the data tape is assigned to x and the next number is assigned to y. In programming, the words input and output

5

are often used as verbs and, if the values of x and y are to be printed, the instructions are written as

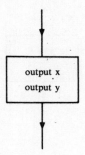

If words are to be printed as part of the output, they are written in blocks as shown:

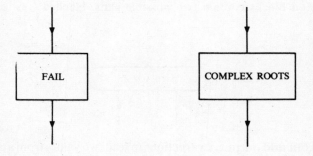

4. Labels. Circles with labels inside are used to mark different points in the diagram.

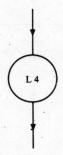

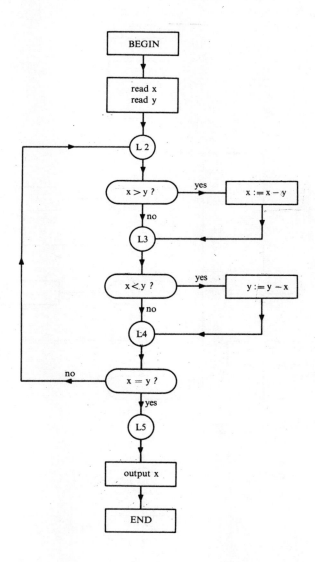

Fig. (i).

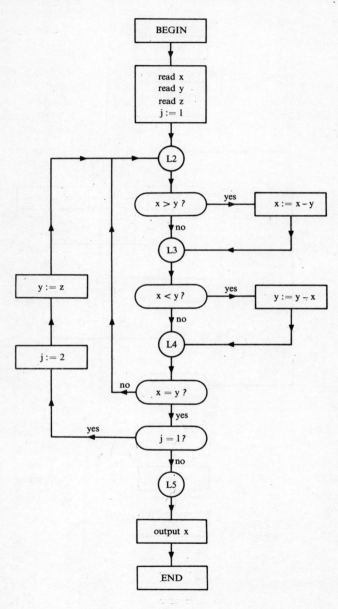

Fig. (ii).

The numbers chosen may often appear arbitrary, but when the equivalent Algol versions are printed, the labels in the flow diagrams are also used to number the lines.

5. Loops. The flow diagram in Fig. (i) makes use of Euclid's method for finding the highest common factor of two integers, x and y, and shows the use of a simple loop. The program starts at L2 and, if x and y are equal at L4, the 'yes' path is taken to L5 where the current value of x is output as the highest common factor. Otherwise a loop is formed and the program is repeated from L2.

Suppose, for instance, that x and y are initially 24 and 32. The highest common factor is 8, and the table below shows the values of x and y as each label is passed.

Label number	x	y
2	24	32
3	24	32
4	24	8
2	24	8
3	16	8
4	16	8
2	16	8
3	8	8
4	8	8
5	8	8

6. Counters. To find the highest common factor of three integers x, y and z, the program could be doubled in length, and then used to find the highest factor common to z and the highest common factor of x and y. Alternatively, a counter j can be used to control the proceedings as in Fig. (ii). This method of reducing the length of a program is much used in computing work.

7. Exit from labels. Fig. (iii) on page 11 shows how an automatic loop can be constructed by means of the statement

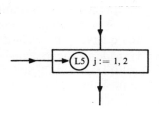

in which j is assigned the value
of 1 and then the value of 2. This
indicates that no change is to be
made in the value of j until label
L5 is reached.

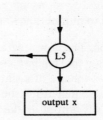

Here there is a choice of paths,
and it is implied that a return
must be made to the assignment
statement unless all the elements
on its right-hand side have been
exhausted. Only in this latter case
can subsequent instructions be
followed.

8. Multiple statements. More advanced programs sometimes require assignment statements in the following form:

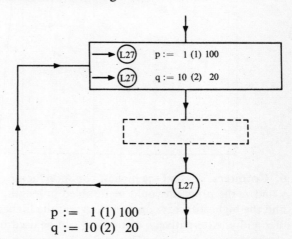

The notation
$$p := 1 \ (1) \ 100$$
$$q := 10 \ (2) \ 20$$

implies that p is assigned all the integers from 1 to 100 and that, for each value of p, q is assigned all the even numbers from 10 to 20. Both p and q retain their current values until label L27 is reached, and at this point a return is made to the assignment statements unless p and q have both reached their maximum values. Each time a return is made to the assignment statements, q is increased by 2 unless its value is already 20, in which case p is increased by 1 and q is returned to the value of 10.

10

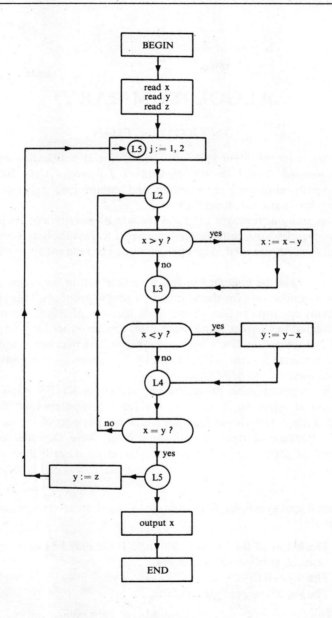

Fig. (iii).

2

ALGOL SUMMARY*

INTRODUCTION

The language of Algol is defined in an official publication entitled *The Revised Report on the Algorithmic Language Algol 60*. It is principally employed in scientific and engineering problems and enjoys international recognition.

The many advantages of Algol include its brevity and its power. Moreover, because conventional mathematical symbols and ordinary English words are used, Algol programs can be read with comparative ease.

To explain the language in full is neither within the scope of this book nor necessary for the writing of a simple program. This chapter contains an introduction to the basic features of Algol. Arrays are considered in Chapter 4 and procedures are introduced in Chapters 6 and 7. Boolean algebra is mentioned only where necessary, and readers are referred to the official manuals for details of more advanced techniques.

The demonstration programs are written in KDF 9 Algol. This version of Algol 60 is for use on KDF 9 computers and includes special input and output facilities which are not part of the language itself. Because of their limited application, these facilities are employed as little as possible and are therefore described in outline only.

Basic Algol symbols. The language of Algol uses four types of basic symbol:

(a) The letters of the English alphabet. These may be lower or upper case, i.e. small or capital letters.
(b) The digits 0 to 9.
(c) The logical values <u>true</u> and <u>false</u>.

---

* This chapter need not be studied in detail at a first reading, and may be considered in conjunction with Chapter 3.

(d) A group of symbols called delimiters. These are:

Arithmetic operators:	$+$ $-$ $\times$ $\div$ $/$ $\uparrow$
Relational operators:	$<$ $\leqslant$ $=$ $\geqslant$ $>$ $\neq$
Logical operators:	<u>and</u> <u>or</u> <u>not</u> <u>eqv</u> <u>imp</u>
Sequential operators:	<u>goto</u> <u>if</u> <u>then</u> <u>else</u> <u>for</u> <u>do</u>
Separators:	, . $_{10}$: ; := *
	<u>step</u> <u>until</u> <u>while</u> <u>comment</u>
Brackets:	( ) [ ] [ ] <u>begin</u> <u>end</u>
Declarators:	<u>own</u> <u>boolean</u> <u>integer</u> <u>real</u> <u>array</u>
	<u>switch</u> <u>procedure</u>
Specificators:	<u>string</u> <u>label</u> <u>value</u>

Those basic Algol symbols which are English words are written in lower case and underlined. It should be noted, however, that Algol symbols do not follow the customary rules of English punctuation. Thus, when examples are printed in the text, the punctuation is that of Algol and not of English.

Decimal numbers. Numbers may be signed or unsigned, e.g.

$$+17.4 \qquad 17.4$$

Integers are written in their normal form:

$$4 \quad -12 \quad 0 \quad +17485$$

Real numbers may include a decimal fraction part. If a decimal point is used, it must be followed by at least one digit:

$$124.3 \quad +0.04 \quad -0.52 \quad .12467 \quad .000$$

Decimal exponents are also permitted. Thus the number 7000 may be written as $7_{10}3$ with the subscript, $_{10}$, printed below the line in small type. The exponent, which must be an integer, may be signed or unsigned, and the preliminary decimal number may be omitted:

$$8_{10}-3 \quad +7.2_{10}2 \quad 0.43_{10}+4 \quad _{10}5 \quad _{10}-12$$

All Algol numbers belong to one of two types, <u>real</u> or <u>integer</u>. If a number contains . or $_{10}$ it is always of type <u>real</u>; otherwise it is of type <u>integer</u>. No other symbols are permitted in an Algol number.

The following examples are NOT acceptable as Algol numbers:

$$45. \quad -12_{10}+2.5 \quad 2_{10} \quad 1,000 \quad 4 \div 2$$

Identifiers. Reference is made to the memory of a computer by giving names or identifiers to variables. Identifiers always start with a letter (upper or lower case) and may be followed by any number of letters or digits:

sum   A14   DETERM II   a5B2

Spaces and line changes are ignored by a computer and may be left between the symbols of an identifier or anywhere in a program to improve the layout.

Identifiers are also used to denote labels which mark particular points in a program. A colon is placed after identifiers used in this manner, e.g.

exit: Fail:

In the programs that follow, labels are frequently used to number the lines, e.g.

L5:   L42:

Declarations. An identifier which is used to refer to a decimal number must first be declared as type <u>real</u> or type <u>integer</u> according to the nature of the number to which it refers:

<u>integer</u> x, Fl;   <u>real</u> A, b, C2, K;

A declaration informs the computer whether a number is to be stored in floating point form or as an integer. It consists of a declarator (see page 13), followed by a list of the identifiers of variables having that type. The identifiers are separated by commas, and the declaration is concluded by a semi-colon.

EXPRESSIONS

Arithmetic expressions. Numbers and variables may be combined by the arithmetic operators

+   −   ×   ÷   /   ↑

to form an arithmetic expression (AE).

The operators / and ÷ both denote division. The former may be used to combine <u>real</u> or <u>integer</u> operands and produces a result of type <u>real</u>. The operator ÷ is often called the integer division sign and may only be used when both operands are of type <u>integer</u>. It

14

gives a result, also of type <u>integer</u>, the value of which is the integer nearer to zero than the value of a/b. Thus

$$27/4 \text{ has a value of } 6.75, \text{ but } 27 \div 4 \text{ has a value of } 6,$$
$$-27/4 \text{ has a value of } -6.75, \text{ but } -27 \div 4 \text{ has a value of } -6.$$

The integer division sign is frequently used in testing an integer for factors, e.g.

$$12 \div 3 \times 3 - 12 = 0$$
$$13 \div 3 \times 3 - 13 \neq 0$$
$$14 \div 3 \times 3 - 14 \neq 0$$

It follows that, in general, j is a factor of k if

$$k \div j \times j - k = 0$$

The operator ↑ is the sign of exponentiation. The base and exponent may have any value providing they do not lead to infinite, indeterminate or imaginary results. If the exponent is of type <u>real</u> the base cannot be negative.

Operators are evaluated in order of occurrence unless the next operator has higher priority.

(1) ↑ has highest priority.
(2) ×, ÷ and / have equal priority after ↑.
(3) + and − have joint lowest priority.

This order of priority may be overruled by the use of brackets:

$$8 \div 4 \div 2 = 1 \qquad 8 \div (4 \div 2) = 4$$
$$2{\uparrow}3{\uparrow}2 = 64 \qquad 2{\uparrow}(3{\uparrow}2) = 512$$
$$2{\uparrow}4 + 3 = 19 \qquad 2{\uparrow}(4 + 3) = 128$$
$$3 + 2{\uparrow}3 = 11 \qquad (3 + 2){\uparrow}3 = 125$$

The operators ↑, ×, + and − give results of type <u>real</u> unless both operands are of type <u>integer</u>.

Adjacent operators are not permitted, nor may one number (or variable) be adjacent to another number (or variable). For example:

$$b^2 - 4a \quad \text{is written as} \quad b{\uparrow}2 - 4 \times a$$
$$a \div -b \quad \text{is written as} \quad a \div (-b) \text{ or as } -a \div b$$

$$\frac{k^2}{(2j - 1)(2j - 2)} \quad \begin{array}{l} \text{is written as} \quad k{\uparrow}2/((2 \times j - 1) \times (2 \times j - 2)) \\ \text{or as} \quad k{\uparrow}2/(2 \times j - 1)/(2 \times j - 2) \end{array}$$

15

The sequential operators

if   then   else

may occur within an arithmetic expression:

if b > 4 then 3 else 2

The result of an AE will always be a single value, and this expression has a value of 3 if b = 5 but a value of 2 if b = 0.

Similarly, when a = 2 and b = 3, the result of

if a > 4 then a↑2 + b↑2 else a + 5

is 7, and the result of

if b < 0 then 2 × b − 3 else if b ⩾ 4 then a + 2 else 4

is 4.

If an AE is unconditional, it is known as a simple arithmetic expression (SAE). An SAE should not be confused with an Algol number. $10\uparrow(-8)$ and $_{10}-8$ both have the same value, but only the latter is an Algol number.

Boolean expressions. The simplest form of boolean expression (BE) uses one of the relational operators

<   ⩽   =   ⩾   >   ≠

Each of these operates on two numbers or variables, and the following are examples of simple boolean expressions:

a > b   x = 0   p ⩽ q

Boolean expressions may be connected by means of logical operators. The most frequently used of these are

and   or

both of which have their natural meanings.

A boolean expression may include an arithmetic expression, but it always possesses one of the two logical values, true or false. If a and b are 3 and 4 respectively, the expression a < b has a value of true. Similarly,

a < b	and	a > 0	has a value of true,
a < b	and	b↑2 < 2 × a	has a value of false,
a < b	or	b↑2 < 2 × a	has a value of true,
a > b	or	b↑2 < 2 × a	has a value of false.

Note than an expression such as a > b or a > c cannot be reduced to a > b or c. Similarly, the expression a > b > c must be written as a > b and b > c, because logical operators may only be used to join two boolean expressions.

STANDARD FUNCTIONS

A number of standard functions may be employed in a program and the arguments of such functions are placed in brackets. An argument may contain any arithmetic expression and is sometimes known as a parameter (see Chapter 6).

abs (AE)	gives the modulus or absolute value of the AE.
sign (AE)	gives one of the values, $+1$, -1, or 0 depending upon the sign of the AE.
sqrt (AE)	gives the positive square root of the AE.
sin (AE)	gives the sine of the AE in radians.
cos (AE)	gives the cosine of the AE in radians.
arc tan (AE)	gives the principal value in radians of $\tan^{-1}$ (AE).
exp (AE)	gives the exponential function of the AE.
ln (AE)	gives the natural logarithm of the AE.
entier (AE)	gives the largest integer not greater than the value of the AE.

Examples:

$$\begin{aligned}
\text{abs}(-8) &\quad \text{has a value of 8,} \\
\text{sign}(-8) &\quad \text{has a value of } -1, \\
\text{arc tan}(\text{sqrt}(3)) &\quad \text{has a value of } \pi/3, \\
\text{entier}(-8/3) &\quad \text{has a value of } -3, \\
\text{entier}(\text{abs}(-8/3)) &\quad \text{has a value of } +2, \\
\ln(\exp(q\uparrow2)) &\quad \text{has a value of } q^2.
\end{aligned}$$

The results of all standard functions are of type real except those from entier (AE) and sign (AE) which are of type integer. Special care must be taken with the square root function because a negative argument will cause a 'route failure'. Similarly, a fail notice will result from ln (AE) unless the arithmetic expression is positive.

A useful application of the entier function is the form

$$\text{entier } (x + 0.5)$$

which gives the nearest integer to x. For example:

$$\begin{aligned}
\text{entier } (3.75 + 0.5) &\quad \text{has a value of 4,} \\
\text{entier } (-9.03 + 0.5) &\quad \text{has a value of } -9.
\end{aligned}$$

STATEMENTS

Instructions to a computer take the form of statements. These are separated from each other by semi-colons and are normally obeyed in order of occurrence.

1. Assignment statements. An assignment statement causes a value to be assigned to a variable. It uses the symbol := which means 'is assigned the value of':

$$z := 5; \quad y := 2 \times z/3; \quad x := z + y;$$
$$x := \underline{\text{if }} y > z \underline{\text{ then }} 2 \underline{\text{ else }} 4;$$

Variables occurring to the right of an assignment statement must previously have a defined value. Thus $x := y + 2$ has no meaning unless y has occurred on the left-hand side of a previous assignment statement.

Providing a variable has a previously defined value, it may occur on both sides of an assignment statement:

$$p := 5; \quad p := 2 \times p; \quad p := -p;$$
$$p := \underline{\text{if }} p > 2 \underline{\text{ then }} p + 2 \underline{\text{ else }} p + 3;$$

Several variables of the same type may be simultaneously assigned a given value:

$$a := b := c := d := 0;$$

If a real number is assigned to an integer variable, it is first rounded to the nearest integer. This rule differs from that of the entier function and integer division sign, and the three methods of assigning real numbers to integer variables are summarized in the table opposite.

2. Goto statements. The normal sequential flow of operations may be interrupted by the insertion of a goto statement in the form

$$\underline{\text{goto }} L;$$

where L is a label. The instruction goto is an Algol basic symbol and may be printed with or without a space.

Simple loops may be performed by means of a goto statement. Thus the instructions contained in Fig. (i) of Chapter 1 may be written as

$$L2: \underline{\text{if }} x > y \underline{\text{ then }} x := x - y;$$
$$\underline{\text{if }} x < y \underline{\text{ then }} y := y - x;$$
$$\underline{\text{if }} x \neq y \underline{\text{ then }} \underline{\text{goto }} L2;$$

Function	Rule	Real value			
		$-1\frac{2}{3}$	$-1\frac{1}{3}$	$1\frac{1}{3}$	$1\frac{2}{3}$
Integer result from division of integers ($\div$)	nearest integer towards zero	-1	-1	1	1
Real number assigned to an integer variable	nearest integer	-2	-1	1	2
entier (AE)	largest integer not greater than the AE	-2	-2	1	1

The table of results shows the values of the functions after assignment to an integer variable.

3. Conditional statements. A statement is said to be conditional when it is preceded by an if clause:

$$\text{if } x > y \text{ then } p := x + y \text{ else } q := x - y;$$

The following points should be noted concerning the use of an if clause:

(a) An if clause consists of the word if followed by a boolean expression followed by the word then:

$$\text{if } x > y \text{ then } p : = x + y;$$

In a statement, the word else is optional, but if else is omitted and the initial condition is not satisfied, the next statement in the sequence will be obeyed.

(b) An if clause may also occur in an arithmetic expression, in which case it must always be followed by the word else:

$$q : = \text{if } x > y \text{ then } p \text{ else } p + 1;$$

It follows that

$$\text{if } a > b \text{ then } p := q;$$

is permissible because it is a conditional statement, but

$$p := \text{if } a > b \text{ then } q;$$

is not permissible because it contains an incorrect arithmetic expression.

19

(c) A statement or expression may contain two or more if clauses, but, in order to prevent ambiguity, the word <u>then</u> may never be followed directly by the word <u>if</u>. Instead, <u>begin</u> . . . <u>end</u> brackets are used to separate the delimiters and to clarify the meaning. The statement,

<u>if</u> 7 < 8 <u>then</u> <u>if</u> 4 < 3 <u>then</u> p := 5 <u>else</u> q := 6;

is ambiguous. It should be written either as

<u>if</u> 7 < 8 <u>then</u> <u>begin</u> <u>if</u> 4 < 3 <u>then</u> p := 5 <u>end</u> <u>else</u> q := 6;

or as

<u>if</u> 7 < 8 <u>then</u> <u>begin</u> <u>if</u> 4 < 3 <u>then</u> p := 5 <u>else</u> q := 6 <u>end</u>;

The first of these three statements is inadmissible, the second would cause no action, and the third would assign a value of 6 to q.

4. <u>For</u> statements. The use of a loop is so important in programming that Algol contains a special facility which enables instructions to be obeyed any number of times. This is the <u>for</u> statement which always contains the sequential operators <u>for</u> and <u>do</u>.

If x has an initial value of 2, the statement

<u>for</u> j := 4, 8, 3 <u>do</u> x := x + j;

will cause x to be increased by 4, 8 and then 3 until it finally attains a value of 17. The table lists the values of the variables j and x under the action of the <u>for</u> statement:

j	.	4	8	3
x	2	6	14	17

j is called the controlled variable, and the numbers assigned to it form the <u>for</u> list elements. These elements may be Algol numbers or exist in one of three distinct forms:

(a) The arithmetic element.
(b) The <u>step-until</u> element.
(c) The <u>while</u> element.

20

(a) The arithmetic element. A <u>for</u> list element may be an arithmetic expression:

$$\underline{\text{for}}\, j := 7 \times y, 3, 4 - y \,\underline{\text{do}}\, z := z \times j;$$

If $y = 2$, this statement will first multiply z by 14, after which a return is made to the <u>for</u> list. The new value of z is multiplied by 3, and the result is then multiplied by 2. If z is initially 1, its final value is 84.

j	·	·14	3	2
z	1	14	42	84

(b) The <u>step-until</u> element. This element provides a simple method by which the number of executions of a <u>for</u> statement can be controlled.

The statement

$$\underline{\text{for}}\, j := 1\, \underline{\text{step}}\, 1\, \underline{\text{until}}\, 5\, \underline{\text{do}}\, b := b \times 2;$$

causes b to be doubled on 5 occasions. The identifier j is used only as a counter and, if b is initially 1, the values of the variables are as listed below:

j	·	1	2	3	4	5
b	1	2	4	8	16	32

The general form of the <u>step-until</u> element is

$$\text{AE} \quad \underline{\text{step}} \quad \text{AE} \quad \underline{\text{until}} \quad \text{AE}$$

Thus, as usual, an arithmetic expression may be inserted in place of an Algol number:

$$\underline{\text{for}}\, j := a + b\, \underline{\text{step}}\, a/3\, \underline{\text{until}}\, 4 \times a + b\, \underline{\text{do}}\, a := 2 \times b;$$
$$\underline{\text{for}}\, k := 2 \times a - b\, \underline{\text{step}}\, -1\, \underline{\text{until}}\, 2 \times a - 3 \times b\, \underline{\text{do}}\, b := j - b;$$

When a <u>step-until</u> element is used, the controlled variable is stepped by intervals until the value of the final arithmetic expression is reached. This final value need not necessarily be equalled but it cannot be passed. Thus, for given starting values of a and b, the following two statements would provide the same results:

$$\underline{for}\ j := 2\ \underline{step}\ 3\ \underline{until}\ \ 8\ \underline{do}\ a := b + j;$$
$$\underline{for}\ j := 2\ \underline{step}\ 3\ \underline{until}\ 10\ \underline{do}\ a := b + j;$$

Similarly, the statements below have the same effect:

$$\underline{for}\ i := 20\ \underline{step}\ -4\ \underline{until}\ 4\ \underline{do}\ a := a + i;$$
$$\underline{for}\ i := 20\ \underline{step}\ -4\ \underline{until}\ 2\ \underline{do}\ a := a + i;$$

(c) **The while element.** Use may also be made of a while element to control the number of times a statement is executed. In the following example, y is initially zero. The <u>for</u> statement assigns to j a constant value of 3, and y is then increased by j. Repeated additions of j are made until y reaches a final value of 18:

$$y := 0;$$
$$\underline{for}\ j := 3\ \underline{while}\ y \leqslant 15\ \underline{do}\ y := y + j;$$

j		3	3	3	3	3	3
y	0	3	6	9	12	15	18

The while element in a <u>for</u> list has the general form

AE <u>while</u> BE

Thus the separator while is preceded by an arithmetic expression or a number, and followed by a boolean expression. In the next example, z is initially 30, and the table of results lists the values of k and z under the action of the following statements.

$$z := 30;$$
$$\underline{for}\ k := 0,\ k + 3\ \underline{while}\ z > 20\ \underline{do}\ z := z - k;$$

k	0	3	6	9
z	30	27	21	12

Note that when k = 6, z is assigned a value of 21. When κ = 9, z is still 21 at the moment the boolean expression is evaluated, and so the statement following <u>do</u> is obeyed once more, thereby giving z a final value of 12.

Notes on <u>for</u> statements

(a) The <u>step-until</u> element may not be combined with the <u>while</u> element in an attempt to form a new element, using the delimiters, <u>step</u> and <u>while</u>. A single <u>for</u> list may, however, contain any number of the standard elements:

<u>for</u> i := 0.5, i × 2 <u>while</u> y < 20, 6, 7 <u>step</u> −1 <u>until</u> 0 <u>do</u>
y := y + i↑2;

(b) Two or more <u>for</u> statements may be combined into a multiple <u>for</u> statement. When two controlled variables are used, the <u>for</u> list of the second variable is exhausted for each element of the first list in turn. The action of a double <u>for</u> statement is shown below:

x := 0;
<u>for</u> j := 1 <u>step</u> 2 <u>until</u> 5 <u>do</u>
<u>for</u> k := 3 <u>step</u> −1 <u>until</u> 0 <u>do</u> x := x + j + k;

j	·	1	1	1	1	3	3	3	3	5	5	5	5
k	·	3	2	1	0	3	2	1	0	3	2	1	0
x	0	4	7·	9	10	16	21	25	28	36	43	49	54

5. Compound statements.

Two or more statements may be grouped together to form a compound statement. This must always be enclosed by the symbols <u>begin</u> . . . <u>end</u> which act as opening and closing brackets. The whole group of statements may be employed whenever a single statement would be allowed. The following example shows the use of two compound statements within a conditional statement:

<u>if</u> a > b <u>then</u> <u>begin</u> p := p + q;
q := 2 × p;
r := q − p
<u>end</u>
<u>else</u> <u>begin</u> p := p − q;
r := 2 × p − q
<u>end</u>;

23

It will be seen that semi-colons are used to separate statements but that no punctuation is placed before either of the end brackets or before else. In general, all statements are terminated by semi-colons unless they are followed immediately by end or else. The vertical lines between the begin . . . end brackets have no Algol significance, but they are often drawn on a printed program to show the underlying structure more clearly.

A compound statement may also be used to form the body of a for statement:

$$
\begin{array}{l}
\underline{\text{for}}\ j := 1\ \underline{\text{step}}\ 1\ \underline{\text{until}}\ n\ \underline{\text{do}} \\
\underline{\text{begin}} \quad a := b + 4; \\
\qquad\quad c := 2 \times b - a \\
\underline{\text{end}};
\end{array}
$$

Exit from a for statement body is permitted, and a jump to an outside label may be made using a goto statement. After a jump, the controlled variable retains its current value unless the for list has been exhausted: the controlled variable has no defined value once a for statement has been completed. It is not permissible to use a goto statement to enter directly into the body of a for statement.

As stated in an earlier section, a label may be used to mark the beginning of any statement, and jumps are made by this means. If, however, it is intended to jump to the end of a for statement body, the word end cannot be labelled because it is not itself a statement. Instead, the preceding statement is terminated by a semi-colon, thus implying that it is no longer the last statement in the body. The label marks an empty or dummy statement and has the same effect as marking the end bracket.

The following examples show how a dummy statement can be formed by the addition of a semi-colon:

$$
\begin{array}{l}
\underline{\text{for}}\ j := 1, 2, 3\ \underline{\text{do}} \\
\underline{\text{begin}} \quad a := 2 \times a; \\
\qquad\quad b := a + j \\
\underline{\text{end}};
\end{array}
\qquad
\begin{array}{l}
\underline{\text{for}}\ j := 1, 2, 3\ \underline{\text{do}} \\
\underline{\text{begin}} \quad a := 2 \times a; \\
\qquad\quad b := a + j; \\
\text{L4:} \\
\underline{\text{end}};
\end{array}
$$

INPUT AND OUTPUT

Each range of computers has its own method of input and output. The following summary describes briefly some of the facilities available on KDF 9.

24

A computer possesses a number of peripheral devices, each of which is given a device number. The most common of these are:

	Device	Device number
Input	Paper tape reader	20
	Card reader	40
Output	Line-printer	30
	Paper tape punch	10

Certain identifiers are reserved for use with these devices in special input and output statements. These include:

open   read   write
close   output   write text

Note that the space between 'write' and 'text' is optional. The single word 'writetext' is used in the programs that follow.

Before data can be read into a computer, the required device must be opened by means of statements such as

open (20);   open (40);

If the results are to be printed, the line-printer is opened. If, however, the results are later to be input as data, the paper tape punch is used. Any device opened during a program must be closed before the end:

close (20);   close (40);

Items of data are input by means of 'read' instructions. The statements

x := read (20);   y := read (20);

will cause the first number on the data tape to be read and assigned to x, after which the next number will be assigned to y. Items of data are always punched using Algol numbers which are separated by semi-colons (or by any other basic symbol that cannot be used in an Algol number).

Layout. Algol numbers can be printed in any desired manner by means of a suitable format expression (FE). The argument of a

25

format expression contains a string of symbols and is enclosed both by string quotes, [ ], and by ordinary brackets, e.g.

format ([ss − nd. dddsddd; cc])

Symbols within string quotes are given a special meaning:

Symbol	Meaning
d	represents a digit
.	defines the position of the decimal point
n	as the first digit replaces initial zeros by spaces
+	prints a sign immediately before the first digit
−	prints a minus sign before the first digit if the expression is negative but leaves a space if the expression is positive
≠	prints a sign in a specified position
s	inserts a space

Additional symbols called terminators may be placed at the end of a format expression. They include the semi-colon, the page-shift (p) and up to three line changes (c), the following combinations being allowed:

;   c   p   cc   ccc   ; p   ; c   ; cc   ; ccc

When the result of a calculation is printed by means of a format expression, it is first rounded to the given number of significant figures. Thus, if a real number is assigned the value of 5/3, it may be printed in a variety of different ways, some of which are shown below:

Format	Result
ddd. ddd	001.667
ndd. ddd	1.667
+nd. ddd	+1.667
−nd. dsd	1.6 7
≠ nd. ddd	+ 1.667

Results may also be printed in floating point form, e.g.

$$\text{format } ([-d.\ dd_{10}+nd;\ ])$$

Only one digit is allowed before the decimal point, and the subscript, $_{10}$, must be followed immediately by a sign and 'nd'. No symbol may follow an exponent except a terminator.

Output. If no special format is needed, a result may be printed in a standard layout which is concluded by a semi-colon and a line change:

$$\text{format } ([\neq d.\ ddddsddddsddds_{10}\neq nd;\ c])$$

If a result is to be printed in the standard layout, it requires an instruction in the form:

$$\text{output (DV, AE);}$$

where DV is a device number and AE is an arithmetic expression.

If a special format is required, the instruction takes the form

$$\text{write (DV, FE, AE);}$$

where FE is a format expression.

A possible 'write' statement is

$$\text{write (30, format ([ddd]), x + y);}$$

and this causes the value of x + y to be printed in the given format, provided that the result lies between 0 and 999.5. Numbers outside this range do not fit the format expression and are automatically printed in the standard layout.

A format expression may be assigned to an identifier of type integer so that subsequent 'write' statements may be shortened:

$$f := \text{format } ([-ndddd.\ dddsddd;\ cc]);$$

$$\text{write (30, f, x1);}$$

$$\text{write (30, f, x2);}$$

$$\text{write (10, f, x1);}$$

$$\text{write (10, f, x2);}$$

27

Headings for results and other explanatory text may be printed by an instruction in the form

writetext (DV, ST);

where ST is a string of Algol basic symbols enclosed by string quotes. Spaces within the string are denoted by asterisks:

writetext (30, [Solution*is**]);

Additional spaces, lines or pages may be left using the symbols s, c or p within additional string quotes:

writetext (10, [xl [4s] x2 [ccc]]);

The string [4s] has the same meaning as [ssss] and as ****. It should be noted that numerals are not normally used inside format expressions. Such expressions may, however, start with up to 15 spaces and, in this case, a numeral may precede the symbol s:

format ([14snd. ddc])

Comments. A convenient feature of Algol is the comment convention by which descriptive notes can be inserted in the program for the benefit of future readers. Comments are ignored by the computer if they are written in Algol basic symbols and if they fall into one of two categories:

(a) Sequences following a semi-colon or begin which commence with the symbol comment and terminate with a semi-colon.

(b) Sequences following end which terminate with end, else or a semi-colon. It follows that neither category of comment may contain a semi-colon and that sequences following end may not contain end or else.

BLOCKS AND PROGRAMS

An Algol block is enclosed by begin . . . end brackets. It may consist of any number of statements but, unlike a compound statement, it includes at least one declaration which must be made at its head. A block may contain inner blocks, and the example opposite shows the structure of a typical program:

28

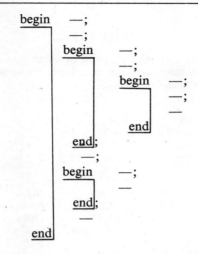

An identifier declared at the head of a block may be employed in that block and also in inner blocks. It may not be used outside the block in which it is declared. An identifier may be redeclared in an inner block, but the program then becomes difficult to follow, and it is preferable to declare all identifiers at the head of the program itself.

KDF 9 Programs. In KDF 9 Algol, the program is separated from its title by an arrow, and from the data by another arrow. The items of data are concluded by a double arrow but, in the absence of data, the program itself is concluded by a double arrow.

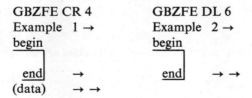

The programs in this book omit the title and initial arrow, and sometimes omit data. Most programs are given in the form of a flow diagram and are accompanied by a version in KDF 9 Algol. On some occasions only a flow diagram is printed in order that a translation may be made as an exercise.

EXERCISES

1. In each of the following sections, some of the sequences of symbols are permissible in Algol and some are not. Mark those that are correctly described by their headings.

(a) Algol numbers

$+22.3$	22.3	±22.3	(-223)
$223.$	$1_{10}-7$	$1._{10}+7$	$_{10}-7$
$10_{10}7$	$_{10}7.0$	3_{10}	$3\times_{10}2$
$3_{10}4\div2$	$3,000$	$-3_{10}2$	$10\uparrow(-2)$

(b) Identifiers

six	true	206	M2.4
Highest factor	XVII	counter	goto
delta	A.E.	p37;	Open
a, or b	sum-diff	Lambda	b + 5
bool.	π	sine	P8C

(c) Arithmetic expressions

$2 - 3(4 - 5)$	$-2.5\uparrow(1/3)$	$5\times(2p - 7)$	W↑D
$b\uparrow2 - 4a\times c$	$(a - 7) \times - 3$	$4\uparrow - 5$	$a + (b)$
$2 \times pi \times \sqrt{l/g}$	$-3\uparrow(-2.5)$	product/two	$12.5 \div 2 - 5$

(d) Standard functions

cos 90	sq. rt (+4)	log (3)	Sin (3/4)
entire (4.7)	ln(sign 3)	sign 0	sqrt (abs(3.5))
ln (0.2)	arc cos (1)	tan⁻¹(1)	àbs(sin(1/2))

2. Mark the expressions which yield numbers of type integer

$3 + 4$	$3/7$	$3.0 - 2.0$	$7 \div 4$
$4\uparrow3$	$(12/4)\uparrow2$	$12 \div 4\uparrow3$	4×2.5
$4_{10}2$	$ln(e\uparrow2)$	$sign(42/13)$	$abs (-4)$
entier (4.3)	sin (0)	$5 \div 2 + 3.5$	sign(sqrt(7))

3. Given $a = 0$, $b = 1$, $c = 2$, $d = -1$, $e = -2$, $f = -3$, calculate the values of the following expressions if all variables are of type integer.

(a) $a + b - c/d$
(b) $c - 2\times b + e \div f$
(c) $e\uparrow2 - 2\times b/c$
(d) $d\uparrow b\uparrow c + 5 \times f \div c \div f$
(e) $(c - d)\uparrow2 - 3\times d \div c$
(f) $-3 \times (b + (-2 \times (7 \times b/(4 - a) - b)))$
(g) $e \times f/(2 \times b + 2) \times (2 \times b + 3)$
(h) if $a > b$ then d else e
(i) if $e > d$ then begin if $c < f$ then a else b end else c
(j) if $a > b$ then d else if $a > c$ then d else e

4. Write out in tabular form the successive values of the variables in the following statements.

(a)   x := 0;
        for i := 7 step −2 until 1 do x := x + i;

(b)   x := 0;
        for i := 7, i − 2 while i ⩾ 1 do x := x + i;

(c)   x := 0;
        i := 9;
   L3: i := i − 2;
        x := x + i;
        if i > 1 then goto L3;

(d)   x := 0;
        for i := 7, 11 ÷ 2, 3, i ÷ 2 do x := x + i;

(e)   sum := 0;
        for i := −2 step 3 until 4 do
        for j := 3, j −2 while j ⩾ −3 do sum := sum + i × j;

3

PRACTICAL PROGRAMMING

RUNNING A PROGRAM

Those without practical experience of computing are liable to assume that running a program is simply a matter of typing the instructions, handing them to an operator and collecting the results. In practice it is not as easy as this, and even a short program may require considerable preparation before it runs successfully.

It will be seen from Fig. (i) that several distinct processes are involved. First, it is worth spending time working on a flow diagram to ensure that all possible eventualities are considered and that the best method is used. The program must then be written and should be given a 'dry run'. This means that the variables should be given comparatively simple values and the program worked through statement by statement, until the end is reached. It is surprising how many errors can be detected in this way. Once the dry run is satisfactory, paper tape or cards may be punched, but before the program can be submitted to the computer for translation into machine code, a suitable title and reference number should be prefixed.

A final check should now be made to ensure that no grammatical errors have been made. The number of possible mistakes is almost unlimited and most of them will cause the program to be rejected. A rejection will be accompanied by a 'fail' notice which will indicate the type of mistake by means of a code number. The offending line will generally be mentioned, and the nature of the error can be ascertained from an error list.

Some mistakes are more common than others. These include:

(a) Omission of semi-colons between statements.
(b) Use of adjacent operators or identifiers.
(c) Failure to declare all the variables used.
(d) Failure to balance each opening bracket by a corresponding closing bracket.

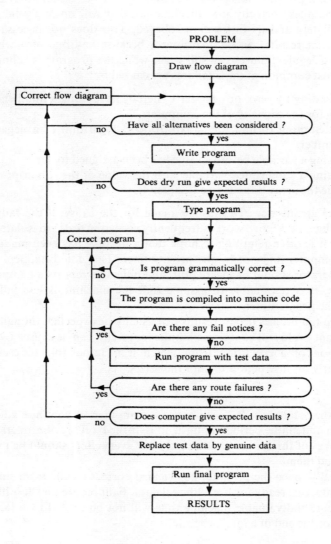

Fig. (i). Running a program.

When a program has been corrected, it may be re-submitted and, if it compiles correctly into machine code, it will normally be run immediately and its instructions obeyed. This does not necessarily mean that results can be guaranteed, because any one of a whole series of 'route' failures can still occur while the program is running. The most common of these failures are caused by:

(a) Dividing by zero, or near-zero, thereby generating a number too big for the capacity of the computer.
(b) Attempting to find the square root or logarithm of a negative number.
(c) Using a variable before a value has been assigned to it.
(d) Using a subscripted variable with the value of the subscript outside the array bounds (see Chapter 4).

Once again, a search must be made for the cause of the failure, and diligent detective work is frequently needed in order to isolate the error. It is often tempting to hope that a mistake has been made by the computer and not by the programmer. This could happen, but the operator should be aware of it, and unless he reports a computer failure, it is reasonable to assume that the machine is less fallible than the programmer.

If no fail notices occur, results can usually be expected, though the program could still contain an infinite loop. This can be caused by the inclusion of a condition from which it is impossible to escape. Similarly an instruction to sum the series

$$1 - 1 + 1 - 1 + \ldots$$

until the total exceeds 2 will cause a computer to continue adding for an indefinite period or until it is turned off by the operator. Mistakes of this nature can be costly and every effort should be made to avoid them.

Finally, once the program has yielded correct results from simple test data, it is ready for use with real data. Failures are now less likely to occur, and valuable machine time will not be wasted by a failure towards the end of a long program.

3.1. SIMPLE PENDULUM

The two versions of the first program illustrate how data may be supplied to a computer, and both calculate the time of swing of a simple pendulum from the formula $t = 2\pi \sqrt{(L/g)}$. The second program differs from the first in that no constant number is assigned to L. Instead, the length of the pendulum is punched on a data tape and is read into the program by the statement

$$L := \text{read (20)};$$

Provision is made for a non-integral length by declaring L as a <u>real</u> variable, and a pendulum of any length may thereby be considered.

```
begin   integer L;              begin   real L, g, pi, t;
        real g, pi, t;                  open (20);
        open (30);                      open (30);
        g := 32.2;                      g := 32.2;
        pi := 3.14;                     pi := 3.14;
        L := 2;                         L := read (20);
        t := 2 × pi × sqrt (L/g);       t := 2 × pi × sqrt (L/g);
        output (30, t);                 output (30, t);
        close (30)                      close (20);
end     → →                             close (30)
                                end     →
                                2;      → →
```

Fig. 3.1. Simple pendulum.

It will be appreciated that the programs could be shortened by use of the statement

$$\text{output (30, 6.28} \times \text{sqrt (L/32.2))};$$

In their present form, however, they illustrate several points of basic grammar.

1. Both programs start with the word <u>begin</u> and finish with the word <u>end</u>.
2. All variables are declared at the beginning of the programs. The variables in the declaration lists are separated by commas, and the lists are terminated by semi-colons. The symbol π is replaced by 'pi' because Greek letters are not used in Algol.

35

3. The declarations are followed by a series of statements which are separated from each other by semi-colons.

4. Values are given to each of the variables by means of assignment statements. No variable is used on the right-hand side of an assignment statement unless a value has previously been assigned to it.

5. The formula for t is written as 2 × pi × sqrt (L/g). Numbers and variables are separated by arithmetic operators, and the square root sign is replaced by sqrt (   ).

6. The result is output on a printer with a device number of 30. This device is opened before the result is due to be printed and closed prior to the end of the program. Similarly, a tape-reader with a device number of 20 is made available for the input of data in the second program.

7. The first program is terminated by a double arrow. In the second program, the data is separated from the program by a single arrow and is concluded by a double arrow.

3.2. SUM OF SQUARES

These two programs show how a <u>for</u> statement may be used as an alternative to a <u>goto</u> statement. Flow diagrams for both methods are given in Fig. 3.2(a), and the Algol versions are printed in Fig. 3.2(b).

The programs find the sum of the squares of the first hundred even integers and tabulate each successive result. They both use the variables j and s, the former to count the number of cycles of the loop, and the latter to store the sum of the squares.

For any value of j, the square of the jth term is added to the variable s which already holds the sum of the previous terms. Thus s is increased to $s + j^2$, and a loop is required to find each new sum. The first example uses a <u>goto</u> statement to cause a jump to the label L4, but in the second, the loop is automatically brought about by means of a <u>for</u> statement. This statement ends at line 7, and so the flow diagram contains the decision label

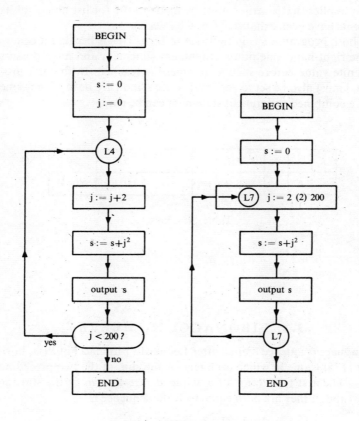

Fig. 3.2(a). Sum of squares.

```
begin  integer j, s;                    begin  integer j, s;
       s := j := 0;                            s := 0;
       open (30);                              open (30);
L4:    j := j + 2;                             for j := 2 step 2 until 200 do
       s := s + j↑2;                           begin  s := s + j↑2;
       output (30, s);                                output (30, s);
       if j < 200 then goto L4;         L7:end;
       close (30)                              close (30)
end    → →                             end    → →
```

Fig. 3.2(b). Sum of squares.

which implies that a return must be made to the for list unless all its elements have been exhausted (see Chapter 1).

In both programs, s is initially set to zero. This is because it occurs on the right-hand side of an assignment statement and must possess a definite value before such a statement is reached. In the first program, j must also be set to zero and, since s and j are both of the same type, a combined assignment statement can be made:

$$s := j := 0;$$

j	0	2	4	6	8	$\cdots$	198	200
s	0	4	20	56	120	$\cdots$	1313400	1353400

Table of values

3.3. FIBONACCI NUMBERS

A Fibonacci sequence (called after Leonardo Fibonacci of Pisa, born *circa* 1175) is one in which each term is the sum of the two preceding terms. The first two terms of a sequence are known as the starting values and, if they are both equal to 1, the sequence is

1   1   2   3   5   8   13 . . .

The flow diagrams in Fig. 3.3 suggest how to calculate the first 50 terms (starting values excluded) of this sequence. Two methods of forming a loop are given. The first diagram shows that a simple jump statement should be employed, and the second implies the use of a for statement. Each diagram assigns a value of 1 to both x and y, but it will be appreciated that, if different sequences are required, suitable starting values can be read from a data tape.

It may appear that 50 Fibonacci numbers are scarcely worthy of calculation, but any attempt to find more will be limited by the capacity of the computer unless the more advanced methods of Chapter 7 are used. KDF 9 has a maximum accuracy of 12 significant figures. It stores real numbers of magnitude up to 2^{128} but integers

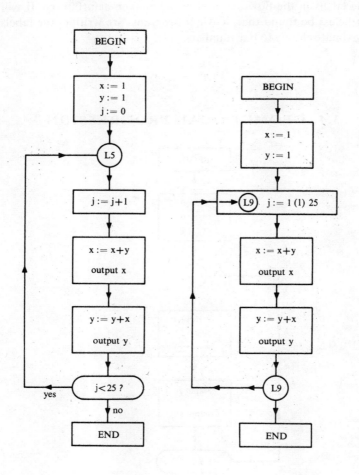

Fig. 3.3. Fibonacci numbers.

up to only 2^{39}. The 50th member of this Fibonacci sequence is 32,951,280,099 and thus approaches the limit of the computer. To exceed 2^{39} would produce a fail notice stating briefly that there had been an 'integer overflow'.

The labels in the flow diagram may be taken as arbitrary. It will nevertheless be found that, if Algol programs are written, the labels approximate closely to line numbers.

3.4. GEOMETRICAL PROGRESSION

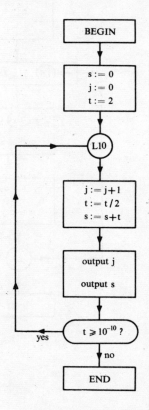

FIG. 3.4(a). Geometrical progression.

```
begin    comment A program to sum the series
         1 + 1/2 + 1/4 + . . . and to print the
         successive results;

         integer j;
         real s, t;
         s := 0;
         j := 0;
         t := 2;
         open (30);

L10:     j := j + 1;
         t := t/2;
         s := s + t;
         write (30, format([ndsss]), j);
         write (30, format([d. dddddsdddddc]), s);
         if t ⩾ ₁₀ − 10 then goto L10;
         close (30)
end      → →
```

Fig. 3.4(b). Geometrical progression.

Program 3.4 finds the sum of the series

$$1 + \frac{1}{2} + \frac{1}{4} + \frac{1}{8} + \ldots + \frac{1}{2^n} + \ldots$$

It prints each successive result and records the number of terms
required. A description is inserted at the head of the program by
means of an Algol comment. All symbols between the word comment
and the next semi-colon are ignored by the computer.

The two series considered previously have both been divergent
and have been ended after a predetermined number of terms. This
program sums a convergent series in which the loop is repeated until
the value of one term is less than a given figure, after which the program automatically ends.

The accuracy of the sum depends on the rounding error introduced
during each loop and also on the sum of the terms omitted from the
series. The latter is called the truncation error and may be calculated

as follows. If n terms have been summed, the truncation error is

$$\frac{1}{2^n} + \frac{1}{2^{n+1}} + \frac{1}{2^{n+2}} + \cdots$$

$$= \frac{1}{2^n}\left[1 + \frac{1}{2} + \frac{1}{4} + \cdots\right]$$

$$= \frac{1}{2^n}\left[\frac{1}{1 - \frac{1}{2}}\right] = \frac{2}{2^n} = \frac{1}{2^{n-1}}$$

The accuracy hoped for in this program is 10 decimal places and terms continue to be added until the value of one is less than 10^{-10} Because of the truncation error, an accuracy of 10 places is not actually achieved, but a good approximation can be obtained in this simple manner. The number of terms needed to reach the given limit is 35 and their sum is found to be 1.9999999999. (The truncation error is $\frac{1}{2^{34}} \doteq 6 \times 10^{-11}$.)

In order to present the intermediate results neatly, the values of the counter, j, and the sum, s, are printed side by side in two columns. Two format expressions are used. The first expression prints the value of j and then leaves three spaces. The second expression prints the value of s correct to 10 decimal places and is terminated by a line change.

3.5. SIN x

Program 3.5 illustrates the use of a 'noise-level' comparison in a program to find $\sin \pi/6$ from the series

$$\sin x = x - \frac{x^3}{3!} + \frac{x^5}{5!} - \frac{x^7}{7!} + \frac{x^9}{9!} - \cdots$$

The result is thereby obtained as accurately as the computer permits, and it can be shown that the truncation error is less than the absolute value of the first of the terms neglected.

The principle of the calculation is to continue adding new terms until the absolute value of one term is equal to or greater than that of its predecessor. Since $\pi/4 < 1$, the absolute value of successive terms decreases. An apparent increase can only occur when two successive terms are smaller than the working accuracy of the computer, in which case rounding errors may result in one term being exceeded by the next.

The method of calculating the terms is important because, if every term is worked out independently, an unnecessary number of multiplications is performed. In order to avoid this, each term is calculated from the one that precedes it. For example, when $j = 2$ the second term is $-x^3/3!$. When $j = 3$ the third term is $x^5/5!$ and can be obtained from the second term on multiplication by $-x^2/5.4$. When $j = 4$ the corresponding multiplying factor is $-x^2/7.6$ and, in general, this factor can be written as $\dfrac{-x^2}{(2j - 1)(2j - 2)}$.

The printed program is restricted to angles between 0 and $\pi/4$ by the use of the logical operator or in the statement

$$\text{if } x > \text{pi}/4 \text{ or } x < 0 \text{ then goto L17;}$$

The evaluation of the sine function is a standard sub-routine in most programming languages and, in practice, more sophisticated methods are employed so that any value of x may be considered.

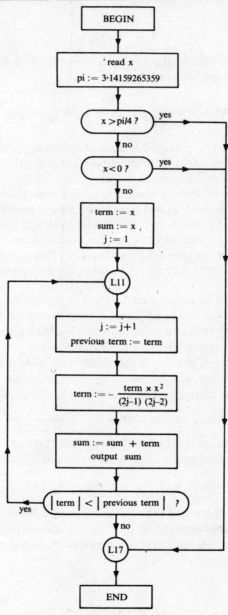

Fig. 3.5(a). Sin x.

```
begin    comment A program to calculate sin x for
         0 ⩽ x ⩽ pi/4, using a noise level calculation and
         printing each successive sum;
         integer j;
         real sum, term, previous term, x, pi;
         open (20);   open (30);
         x := read (20);
         pi := 3.14159265359;
         if x > pi/4 or x < 0 then goto L17;
         term := sum := x;   j := 1;
L11:     j := j + 1;
         previous term := term;
         term := −(term × x↑2)/((2×j − 1) × (2×j − 2));
         sum := sum + term;
         output (30, sum);
         if abs (term) < abs (previous term) then goto L11;
L17:     close (20);   close (30)
end      →
0.523598775598; → →
```

Fig. 3.5(b). Sin x.

3.6. EXPONENTIAL SERIES

The flow diagram in Fig. 3.6 suggests a method to find e^x, for a positive value of x, using the series

$$e^x = 1 + x + \frac{x^2}{2!} + \frac{x^3}{3!} + \frac{x^4}{4!} + \ldots + \frac{x^n}{n!} + \ldots$$

The value of x may be such that the values of successive early terms initially increase. For this reason, a noise-level comparison is made between the sums of the terms and not between the terms themselves. If the series is terminated after n terms, it can be proved that the truncation error is not greater than $\frac{e^x}{(n + 1)!}$.

The method of calculating each term from its predecessor is similar to that of program 3.5. The Algol version of this program is therefore

45

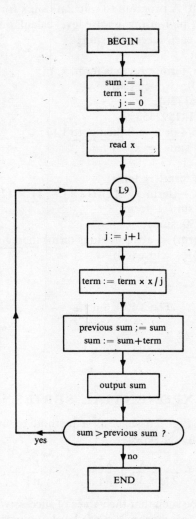

Fig. 3.6. Exponential series.

omitted, and the problem of terminating the series when x is negative is set at the end of the chapter.

3.7. HERO'S FORMULA

This program uses the formula

$$\Delta = \sqrt{s\,(s - a)\,(s - b)\,(s - c)}$$

to find the area of a triangle when the lengths of all three sides are known. The lengths of the sides a and b are supplied as data, and the area is calculated for all possible integral values of the third side, c.

Both in the program and in the flow diagram, use is made of the entier function to convert real numbers to the nearest integer in the direction of zero. The program also employs two format expressions, and these are assigned to two variables, f1 and f2, which are declared as type <u>integer</u>. Both expressions specify that results are to be printed in floating point form.

The values of c to be considered are those integral values for which a triangle can be constructed. They are given by the conditions

$$|a - b| < c < |a + b| \qquad (1)$$

and the limits of c are calculated in line 9 by means of the assignment statements

$$p := 1 + \text{entier (abs } (a - b));$$
$$q := \text{entier } (a + b);$$

These values of p and q satisfy the conditions

$$|a - b| < p \text{ and } q < |a + b|$$

which may be deduced from (1).

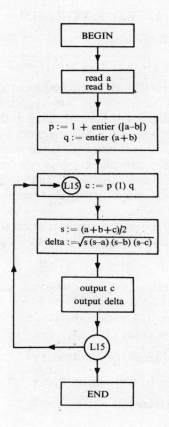

Fig. 3.7(a). Hero.

```
begin      comment  A program to find the areas of triangles from the
           lengths of their sides;

           integer c, p, q, f1, f2;
           real a, b, s, delta;
           f1 := format ([d. dddsddd₁₀ − nd]);
           f2 := format ([sssd. dddsddd₁₀ − ndc]);
           open (20);  open (30);
           a := read (20);  b := read (20);
           p := 1 + entier (abs(a − b));   q := entier (a + b);

           for c := p step 1 until q do
           begin    s := (a + b + c)/2;
                    delta := sqrt (s×(s − a)×(s − b)×(s − c));
                    write (30, f1, c);
                    write (30, f2, delta);
     L15: end;
           close (20);    close (30)
end
45.73; 51.28; → →
```

Fig. 3.7(b). Hero.

PROBLEMS

1. Write programs to find the sum of the following series and to record the number of terms required:

(a) $1 + \left(\dfrac{1}{2}\right)^2 + \left(\dfrac{1}{3}\right)^3 + \left(\dfrac{1}{4}\right)^4 + \ldots$ to 10 decimal places,

(b) $1 - \dfrac{1}{2} + \dfrac{1}{3} - \dfrac{1}{4} + \dfrac{1}{5} - \ldots$ to 3 decimal places,

(c) $\dfrac{1}{2!} + \dfrac{2}{3!} + \dfrac{3}{4!} + \ldots$ as accurately as possible.

2. Write programs to evaluate each of the following functions from a suitable series:
(a) $\cos x$ for $0 \leqslant x \leqslant \pi/4$,
(b) e^x for $|x| < 10$,
(c) $\sin x$ for all values of x.

49

3. By considering the sequence $x = 1, \frac{1}{2}, \frac{1}{4}, \ldots, (\frac{1}{2})^n, \ldots$, find the number of terms required before

$$\frac{\tan x}{x} - \frac{\sin x}{x} < 10^{-6}.$$

Print the successive results in five columns with headings as shown:

$$n \qquad x \qquad \tan(x)/x \qquad \sin(x)/x \qquad \text{difference}$$

4. Consider the Fibonacci series

$$a_1, a_2, a_3, \ldots, a_{50}$$

with starting values a_1 and a_2 supplied as data.
Write a program to calculate
(a) the terms of the series,
(b) the ratio between successive terms,
(c) the ratio between alternate terms,
(d) the differences between these ratios.
Print the results in four columns as shown:

$$a_3 \qquad a_3/a_2 \qquad a_3/a_1 \qquad a_3/a_1 - a_3/a_2$$

$$a_{50} \qquad a_{50}/a_{49} \qquad a_{50}/a_{48} \qquad a_{50}/a_{48} - a_{50}/a_{49}$$

Hence show that $\lim\limits_{n\to\infty} \dfrac{a_n}{a_{n-1}} \to \frac{1}{2}(\sqrt{5} + 1)$

and $\lim\limits_{n\to\infty} \left(\dfrac{a_n}{a_{n-2}} - \dfrac{a_n}{a_{n-1}}\right) \to 1$

5. The lengths of two sides of a triangle are supplied as data and the angle θ between them is an integral number of degrees ($0 < \theta < 180$).
Find for each value of θ:
(a) the length of the third side of the triangle,
(b) the other two angles of the triangle.

6. The lengths of two sides of a triangle are supplied as data, and the third side is an integer. Consider all possible triangles and find each of the three angles by an independent calculation. Find also the difference between each angle sum and 180 degrees.

4

INTEGERS AND ARRAYS

This chapter is concerned with some of the properties of integers, and the demonstration programs consider binomial coefficients, Pythagorean triads, primes and binary digits. Arrays are also introduced and are used to handle large sets of numbers. In computing, arrays are of great importance because they can lead to a marked reduction in the length of a program.

ARRAYS

An array in Algol corresponds to a matrix in mathematics and is used to store a set of numbers. Consider the equation

$$4x^3 + 3x^2 + 8x - 5 = 0 \qquad (1)$$

The coefficients of the left-hand side can be written in matrix form as

$$(4 \quad 3 \quad 8 \quad -5)$$

and in Algol, this matrix is termed a one-dimensional array. Similarly, the equation

$$9.5x^3 - 2x^2 + 3 = 0 \qquad (2)$$

can have its coefficients stored as

$$(9.5 \quad -2 \quad 0 \quad 3).$$

An array is referred to by a single name or identifier and, by means of subscripts, this name may also refer to the individual elements of the array. In the examples above, the two arrays may be given the identifiers A and B, and their elements numbered as shown:

Coefficients   Arrays A and B

4	3	8	5
9.5	−2	0	3

A[1]	A[2]	A[3]	A[4]
B[1]	B[2]	B[3]	B[4]

A[1], B[3], . . . are known as subscripted variables, and the figures 1, 2, 3, 4 are called the subscripts. Using this system, the coefficients 8 and -2 are stored in cells which are identified by the subscripted variables A[3] and B[2]. These variables may be employed in a program wherever a normal variable can be used:

$$p := 5 + A[3]; \quad q := A[4] + B[2]; \quad A[1] := 4 \times p;$$

Instead of storing the coefficients in two one-dimensional arrays, a single two-dimensional array could be used. If this is done, and the new array is given the identifier C, each coefficient may be described by means of two subscripts, the first denoting its row and the second its column:

Coefficients   Array C

4	3	8	5
9.5	−2	0	3

C[1,1]	C[1,2]	C[1,3]	C[1,4]
C[2,1]	C[2,2]	C[2,3]	C[2,4]

The coefficients 8 and -2 are now represented by the subscripted variables C[1, 3] and C[2, 2] which, as before, may be used in the same way as other variables:

$$C[2, 2] := \underline{if}\ A[1] > B[3]\ \underline{then}\ 1\ \underline{else}\ B[2];$$
$$A[4] := C[1, 3] \times 6/C[2, 4]{\uparrow}3;$$
$$C[p, q] := C[2, B[A[1]]];$$

The last example is unlikely to occur in practice, but, taking values from the tables above,

$$A[1] = 4, \quad B[4] = 3, \quad C[2, 3] = 0$$

and hence

$$\begin{aligned} C[2, B[A[1]]] &= C[2, B[4]] \\ &= C[2, 3] \\ &= 0. \end{aligned}$$

Before an array can be used, it must be declared at the head of a block. The type of the array is normally included, but, if it is omitted, the array is understood to be of type <u>real</u>. The declaration also tells the computer the number of cells that the array will need, and, if the array has one dimension, this number is indicated by means of a bound-pair placed immediately after the array identifier. If the array is two-dimensional, separate bound-pairs form a bound-pair-list and indicate the number of rows and columns:

$$\underline{\text{integer}} \;\; \underline{\text{array}} \;\; A[1:4];$$
$$\underline{\text{real}} \;\; \underline{\text{array}} \;\; B[1:4];$$
$$\underline{\text{array}} \;\; C[1:2, 1:4];$$

A bound-pair specifies not only the number of cells needed, but also the limits on the values of the subscripts.

There is no necessity to use any particular numbering system. If, for instance, equations (1) and (2) had been the second and fourth of a set of four equations, they might have been stored in the array D as shown:

Coefficients

•	•	•	•
4	3	8	5
•	•	•	•
9.5	–2	0	3

Array D

•	•	•	•
D[2,13]	D[2,14]	D[2,15]	D[2,16]
•	•	•	•
D[4,13]	D[4,14]	D[4,15]	D[4,16]

The first subscript still indicates the equation, but the second is chosen arbitrarily. The declaration of an array to hold these four equations is

$$\underline{\text{array}} \;\; D \; [1:4, 13:16];$$

The following points concerning the declaration of arrays should be noted:

(a) The lower bound of each pair must be stated before the upper bound.

(b) Any number of arrays of the same type may be declared together:

$$\underline{\text{array}} \;\; X[1:4], Y[1:10], Z[-4:0, 7:12];$$
$$\underline{\text{integer}} \;\; \underline{\text{array}} \;\; Y2[4:7, 1:10], \text{store} [0:6, 1:10];$$

(c) Arrays of the same type, with the same number of dimensions, and with the same bounds may be declared together without repeating the bounds:

$$\underline{\text{array}} \;\; p1, p2, q[1:4], r, t[0:100, -12:-4];$$
$$\underline{\text{integer}} \;\; \underline{\text{array}} \;\; E, f[-7:8], G[1:10, 1:10, 1:10];$$

(d) A variable may be used in a bound-pair-list, but it must have a defined value at the moment the array is declared. Since declarations must precede all other statements in a block, it is impossible for a variable to receive a value until after the declarations at the head of the program have been completed. For this reason, an array with a variable in the bound-pair-list is always declared at the head of an inner block. The variable must, however, be previously assigned a value in an outer block.

```
begin    integer m, n;
         array E[0: 4];
         open (20);
         m := read (20);    n := read (20);
         begin    array F[m: 2 × n];
```

(e) Once an array has been declared and the bound-pair-list evaluated, all array elements must have subscripts within the declared bounds. If an array is declared as

$$\text{array } B[-3: 7];$$

it is meaningless to refer to $B[-5]$ or $B[10]$, either of which will cause the fail notice, 'outside array bounds'.

EXERCISES

In these exercises use arrays declared as follows:

$$\text{integer array} \quad A [1: 3, 1: 3];$$
$$\text{array} \quad B, C [1: 3, 1: 3];$$

1. Verify that the given assignment statements fill the array A as shown:

-1	2	3
4	4	2
0	0	-2

Array A

(a) A[1, 1] := −5 ÷ 3;
(b) A[1, 4 ÷ 2] := 5/3;
(c) A[1, 3] := abs (A[1, 1]) + A[1, 2];
(d) A[2, 1] := A[2, 5 ÷ 2] := A[1, 2]↑2;
(e) A[2, 3] := A[1, A[2, 2] ÷ A[1, 2]];
(f) A[3, 1] := A[1, 3] −3 × abs (A[1, 1]);
(g) A[3, 2] := if A[1, 2] = 0 then A[2, 2] else A[3, 1];
(h) A[3, A[1, 3]] := A[4−A[1, 2], 2] − 6

2. Using the values of array A in question 1, follow the action of the statements below and hence fill the array B as shown:

Array B

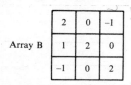

(a) <u>for</u> j : = 1 <u>step</u> 1 <u>until</u> 3 <u>do</u> B[1, j] : = A[3, j];
(b) <u>for</u> j : = 1, $\overline{2, 3}$, <u>do</u> B[2, j] : = A[j, 2];
(c) <u>for</u> i : = 1 <u>step</u> 1 <u>until</u> 3 <u>do</u> B[i, i] : = 4;
(d) <u>for</u> j : = 1, $\overline{j + 1}$ <u>while</u> j < 3 <u>do</u> B[3, j] : = B[j, 3];
(e) <u>for</u> i : = 1 <u>step</u> 1 <u>until</u> 3 <u>do</u>
    <u>for</u> j : = 1 <u>step</u> 1 <u>until</u> 3 <u>do</u> B[i, j] : = B[i, j]/2;

3. Write Algol statements to carry out the following instructions:

(a) Fill the array C with elements that are twice the corresponding elements of the array A (see question 1).
(b) Transpose the rows and the columns of C (i.e. row k becomes column k and vice versa).
(c) Interchange columns 1 and 3.
(d) Interchange rows 1 and 3.
(e) Replace row 3 by itself less row 2.
(f) Find the product of the leading diagonal terms (i.e. the product of C[1, 1], C[2, 2] and C[3, 3]).

Carry out the above instructions and show that the final product is 192.

4.1. BINOMIAL COEFFICIENTS

Binomial coefficients are defined as the coefficients of x in the expansion of $(1 + x)^n$, where n is a positive integer. Program 4.1 calculates these coefficients for integral values of n from 1 to 32, and demonstrates the use of an array with a variable bound.

For each value of n, the array C stores the coefficients of x. It requires n + 1 elements and is declared as

<u>integer</u> <u>array</u> C[0: n];

Since n is a variable, it must have a defined value at the time array C is declared. n is therefore assigned a value in the first block and, because declarations are placed at the head of a block, a new block is constructed and C is declared at its head. The array will initially contain two elements, but it will increase in size at each declaration until it finally contains thirty-three elements.

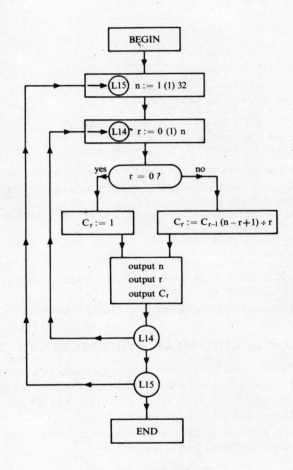

Fig. 4.1(a). Binomial coefficients.

It is possible to print the coefficients directly without employing subscripted variables, but use of an array enables a complete set of coefficients to be stored in the computer at one time.

```
begin        comment A program to calculate binomial coefficients and
             to illustrate the use of an array with a variable bound;
             integer n, r, f1, f2;
             f1 := format ([ndss]);   f2 := format ([nddddddddc]);
             open (30);

             for n := 1 step 1 until 32 do
             begin     integer array C[0: n];
                       for r := 0 step 1 until n do
                       begin   C[r] := if r = 0 then 1
                                    else C[r − 1] × (n − r + 1) ÷ r;
                               write (30, f1, n);
                               write (30, f1, r);
                               write (30, f2, C[r]);
                  L14: end;
             L15: end;
             close (30)
end          → →
```

Fig. 4.1(b). Binomial coefficients.

The program uses a double for statement and, for each value of n, every value of the counter r is considered. If $r = 0$, C[r] is assigned the value of 1, but otherwise the value of C[r] is calculated from C[r − 1] using the recurrence relation

$$C[r] = C[r − 1] \times \frac{n − r + 1}{r}$$

Allowance is made for both cases in the arithmetic expression

C[r] := if r = 0 then 1 else C[r − 1] × (n − r + 1) ÷ r;

Care must be taken not to reverse the order in the expression following else. If C[r − 1] ÷ r × (n − r + 1) is written, the division will be performed first and, since r is not necessarily a factor of C[r − 1], any correction resulting from the integer division sign will cause an error.

The format expressions f1 and f2 are employed to print n, r and C[r] in three columns. The largest coefficient to be printed is $^{32}C_{16}$ (= 601,080,390), and provision is therefore made for printing integers of up to nine digits.

4.2. PYTHAGOREAN TRIADS

A Pythagorean triad is formed when three integers a, b and c satisfy the relation $a^2 + b^2 = c^2$. The most elementary Pythagorean triad is 3, 4, 5 and this may be called 'basic' because its elements contain no common factor. Thus 6, 8, 10 and 15, 36, 39 are multiples of other triads and are not themselves basic.

Flow diagram 4.2 gives a trial and error method to find basic Pythagorean triads in which each integer is less than 100. A double for statement considers integral values of a and b and tests whether $\sqrt{a^2 + b^2}$ is also an integer. To avoid duplication of the results, b is initially assigned a value of a + 1. By this means, no triad such as 65, 72, 97 is printed again as 72, 65, 97.

The square root of $a^2 + b^2$ is assigned to the real variable d and also to the integer variable c which takes the value of the nearest integer. If c and d are equal, a Pythagorean triad has been found.

The values of a and b are now assigned to the variables x and y so that the former values are preserved during the calculation of the highest common factor. This is found using Euclid's method as explained in Chapter 1. Since multiples of basic triads are not required, the results are only printed if the highest common factor is 1.

4.3. TRIADS (DIRECT METHOD)

The previous program found Pythagorean triads by trial and error. It was unnecessarily lengthy and program 4.3 suggests a direct method by which the same results can be obtained in about one-fiftieth of the time.

It can be shown that the sides of an integer right-angled triangle can always be expressed by a triad in the form

$$n^2 - k^2, 2nk, n^2 + k^2.$$

If n and k have no common factor, these terms give a basic triad, though it is always necessary to divide by 2 if n and k are both odd. For example, if n = 5 and k = 3, the triad is 16, 30, 34 and, after

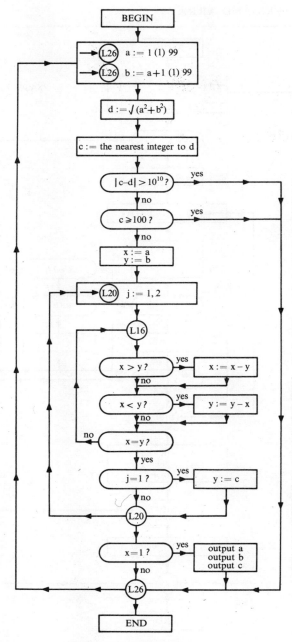

Fig. 4.2. Pythagorean triads.

59

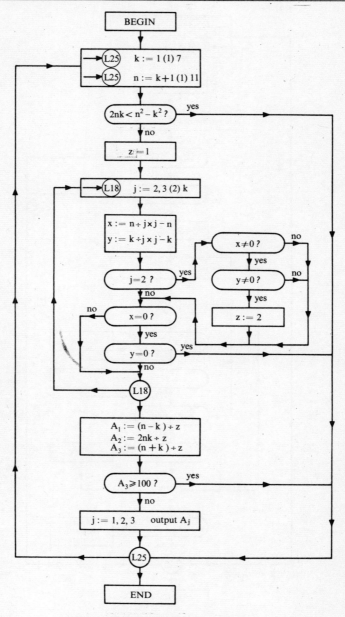

Fig. 4.3(a). Triads (direct method).

division by 2, the basic triad 8, 15, 17 is formed. If n = 3 and k = 2, the basic triad 5, 12, 13 is formed directly.

Analysis of the problem shows that n and k must be considered up to 11 and 7 respectively. Also, since duplicated results are not required, triads are only investigated if $n > k$ and $2nk > n^2 - k^2$.

The statements
$$x := n \div j \times j - n;$$
$$y := k \div j \times j - k;$$

test n and k to find if j is a factor (see page 15). If x and y are both zero, j is a common factor and the triad is rejected. Also, if x and

```
begin      comment A program to generate triads directly;
           integer j, k, n, x, y, z, f;
           integer array A[1 : 3];
           f := format ([ndss]);
           open (30);

           for k := 1 step 1 until 7 do
           for n := k + 1 step 1 until 11 do
           begin   if 2 × n × k < n↑2 − k↑2 then goto L25;
                   z := 1;
                   for j := 2, 3 step 2 until k do
                   begin    x := n ÷ j × j − n;
                            y := k ÷ j × j − k;
                            if j = 2 then
                            begin   if x ≠ 0 and y ≠ 0
                                    then z := 2
                            end;
                            if x = 0 and y = 0 then goto L25;
           L18: end;
                   A[1] := (n↑2 − k↑2) ÷ z;
                   A[2] := 2 × n × k ÷ z;
                   A[3] := (n↑2 + k↑2) ÷ z;
                   if A[3] ⩾ 100 then goto L25;
                   for j := 1, 2, 3 do write (30, f, A[j]);
                   writetext (30, [[c]]);
           L25: end;
           close (30)
end        → →
```

Fig. 4.3(b). Triads (direct method).

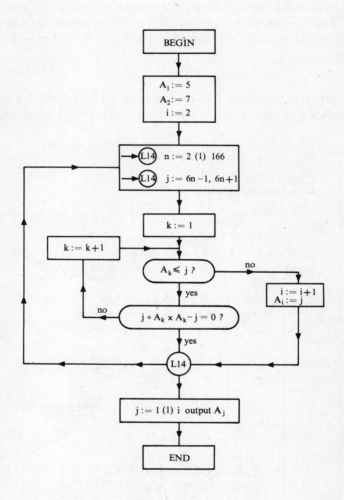

Fig. 4.4(a). Prime numbers.

y are both non-zero when $j = 2$, it follows that n and k are both odd. The variable z is then given a value of 2 and is used later to halve the elements of array A in which the triad is stored.

Results

Time by method 4.2—2 min. 28 sec.
Time by method 4.3—0 min. 3 sec.

3	5	7	8	9	11	12	13	16	20	28	33	36	39	48	65
4	12	24	15	40	60	35	84	63	21	45	56	77	80	55	72
5	13	25	17	41	61	37	85	65	29	53	65	85	89	73	97

4.4. PRIME NUMBERS

A table of prime numbers may be constructed by means of a device

```
begin    comment A program to find all three-digit prime numbers;
         integer i, j, k, n, f;
         integer array A[1 : 200];
         open (30);
         f := format ([nddc]);
         A[1] := 5;   A[2] := 7;
         i := 2;
         for n := 2 step 1 until 166 do
         for j := 6 × n − 1, 6 × n + 1 do
         begin   for k := 1, k + 1 while A[k] < sqrt(j) do
                         if j ÷ A[k] × A[k] − j = 0 then goto L14;
                 i := i + 1;
                 A[i] := j;
    L14: end   of storage of prime numbers in array A;

         for j := 1 step 1 until i do write (30, f, A[j]);
         close (30)
end    → →
```

Fig. 4.4(b). Prime numbers.

known as the Sieve of Eratosthenes. If all the integers are written
down, and the sequences

$$
\begin{array}{cccc}
4, & 6, & 8, & 10 \quad \ldots\ldots \\
9, & 12, & 15, & 18 \quad \ldots\ldots \\
25, & 30, & 35, & 40 \quad \ldots\ldots
\end{array}
$$

. . . .

are struck out, the remaining numbers are primes.

Program 4.4 finds all the three-digit prime numbers but, instead of
employing Eratosthenes' sieve, it uses the fact that every prime greater
than 3 is of the form $6n \pm 1$.

The array A is used to store the primes as they are formed. Their
number is not known in advance, and so A is declared with an upper
bound chosen arbitrarily, but sufficiently big to leave a comfortable
reserve of storage space:

integer array A[1 : 200];

No number of the form $6n \pm 1$ is divisible by 2 or 3. The program
is therefore started by storing the next two primes, 5 and 7, in the
array A. As the integer counter n increases from 2 to 166, values of
$6n \pm 1$ are assigned to j, and these values of j are then divided by
smaller known primes.

Since all non-prime numbers contain a factor less than or equal
to their square root, it is only necessary to divide j by primes less
than or equal to $\sqrt{j}$. The flow diagram illustrates the method, and a
program based on it would normally use a simple goto statement.
Program 4.4(b), however, shows how a while element of a for statement can have the same effect.

If j is found to have a factor, its current value is discarded. If there
is no factor, j must itself be a prime, and its value is transferred to
array A. The number of primes in this array is recorded by the
counter, i, which is increased by 1 whenever a prime is found. Note
that the end of storage of primes in A is recorded in the program by
means of an 'end comment'. The symbols between end and the next
semi-colon are ignored by the computer. In line 10, sqrt(j) is evaluated afresh in each cycle of the loop. This lengthy process can be
avoided by the method of program 4.5.

The results from this program show that the number of primes
less than 1000 is 168. Of these, 46 are less than 200, and each suc-
ceeding set of 100 integers contains between 14 and 17 additional
primes.

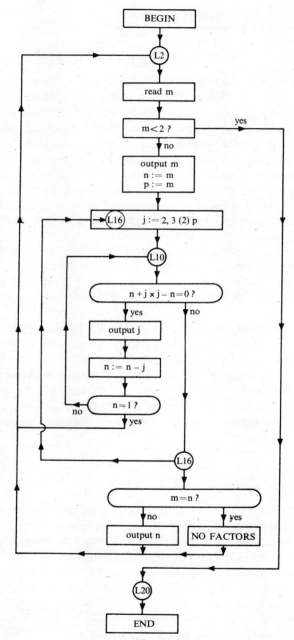

Fig. 4.5. Prime factors.

4.5. PRIME FACTORS

The flow diagram in Fig. 4.5 finds the prime factors of a set of integers. The number of integers is not known in advance, and so the computer is instructed to factorize any integer greater than or equal to 2. Thus the program can be terminated by providing an integer less than 2 as the last piece of data, and there is no limitation on the number of integers that can be punched on the data tape.

Each integer is assigned to both m and n. The former retains its value throughout the program, but the latter is factorized by trial and error and is reduced in value as it is divided by each factor in turn. The integer variable p is assigned the value of $\sqrt{m}$ and automatically takes the value of the nearest integer.

The controlled variable, j, is assigned the values of 2, 3, 5, 7, . . . , p, and, for each value of j, n is tested for divisibility. If a factor is found, the current value of j is printed and n is replaced by $n \div j$. A secondary loop is then formed so that repeated factors are not missed.

It is unnecessary to test m for divisibility by all smaller integers, and the time of the program is halved by ignoring even numbers greater than 2. A far more significant time reduction is made, however, by considering j up to $\sqrt{m}$ only. It is possible that one factor of m is greater than $\sqrt{m}$ (e.g. $78 = 2 \times 3 \times 13$), and thus the final value of n is printed as a factor unless $n = 1$. If the integer under consideration is a prime, m and n remain equal and the words 'no factors' are printed.

4.6. NIM

In a game of Nim played with piles of matches, two players move alternately and take any number of matches from one pile, the winner taking the last match. If a player can set up a winning position, he cannot lose unless he makes a mistake in a subsequent move.

A winning position can be found by expressing the number of matches in each pile in binary form. The binary digits or 'bits' are stored in an array and, if the columns of the array are added separately and each sum is even, this is a winning position. For example, three piles containing 1, 8 and 9 matches can be represented by binary digits as shown. The sum of every column is even, and a player leaving this combination of matches is in a winning position.

Number of matches	Binary digits
1	0 0 0 1
8	1 0 0 0
9	1 0 0 1
	2 0 0 2

An opponent receiving these piles can take matches from one pile only. Since no binary digit can change by more than 1, the sum of at least one column must become odd when the opponent moves. Thus the opponent cannot now create a winning position for himself, and the original player is free to do so at his next move. Winning positions can therefore be left only once in two moves, and the first player to leave one can always win.

This program considers three piles of up to 15 matches. The binary digits of the numbers 1 to 15 are stored in a two-dimensional array and are calculated from the corresponding decimal numbers by subtraction of the integers 8, 4, 2 and 1. A subtraction is performed if it leads to a positive or zero result, and a 1 is then placed in the array A; otherwise a 0 is stored.

The array declaration is

<u>integer</u> <u>array</u> A[1: 15, 0: 3];

This shows that the binary numbers are stored in rows numbered from 1 to 15 and in columns headed by the integers from 0 to 3. A decimal number such as 13 can be written in binary form as $1.2^3 + 1.2^2 + 0.2^1 + 1.2^0$, and thus the column headings are the powers to which the base 2 is raised. These powers are used as array subscripts, and the binary digits of 13 are stored as shown:

Column headings (powers of 2)	3	2	1	0
Subscripted variables	A[13,3]	A[13,2]	A[13,1]	A[13,0]
Binary equivalent of 13	1	1	0	1

Although the array bounds of the columns must be given as 0: 3 with the lower bound first, the array may be imagined as holding the

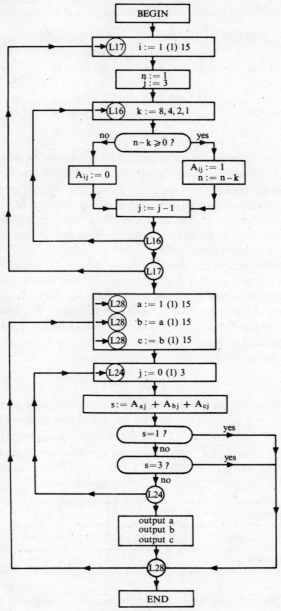

Fig. 4.6(a). Nim.

```
begin    comment A program to output winning positions in a game
         of Nim with three piles of up to 15 matches in each;
         integer a, b, c, i, j, k, n, s, f1, f2;
         integer array A[1 : 15, 0 : 3];
         f1 := format ([ssnd]);   f2 := format ([5sndc]);
         open (30);

         for i := 1 step 1 until 15 do
         begin    n := 1;  j := 3;
                  for k := 8, 4, 2, 1 do
                  begin    if n − k ⩾ 0 then
                           begin    A[i, j] := 1;
                                    n := n − k
                           end
                           else A[i, j] := 0;
                           j := j − 1;
         L16: end;
L17: end;

         for a := 1 step 1 until 15 do
         for b := a step 1 until 15 do
         for c := b step 1 until 15 do
         begin    for j := 0 step 1 until 3 do
                  begin    s := A[a, j] + A[b ,j] + A[c, j];
                           if s = 1 or s = 3 then goto L28;
         L24: end;
                  write (30, f1, a);
                  write (30, f1, b);
                  write (30, f2, c);
L28: end;
         close (30)
end      → →
```

Fig. 4.6(b). Nim.

'bits' in their usual order. The exact location of 'bits' within the computer is of little concern to the programmer.

4.7. BINARY FRACTIONS

The flow diagram shows how the decimal fractions

$$0.99, 0.98, 0.97, \ldots, 0.01$$

may be converted into binary fractions. The value of each decimal is multiplied by 2 on ten successive occasions so that 10 binary places are obtained. The integer part after multiplication becomes a binary digit, and only the remainder is used for further multiplication. The example below shows how 0.77_{10} may be converted to 0.110_2 to three binary places:

$$
\begin{array}{r}
0.77 \\
2 \\
\hline
(1).54 \\
2 \\
\hline
(1).08 \\
2 \\
\hline
(0).16 \\
\hline
\end{array}
$$

A one-dimensional integer array, $A[-10:1]$, is employed, and the result of each conversion is printed immediately so that A can be used repeatedly. As in the previous program, the array subscripts are the powers to which the base 2 is raised.

The results of program 4.7 may be printed in two columns as shown:

Decimal Fraction	Binary Fraction
0.99	0.1111110101
.	.
.	.
0.01	0.0000001010

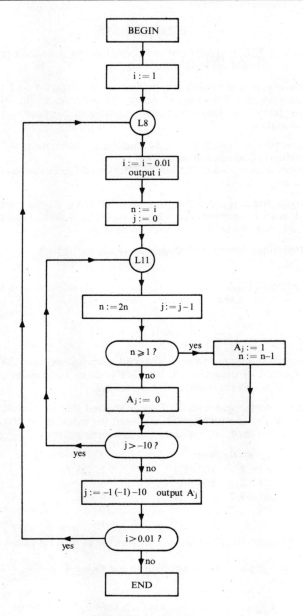

Fig. 4.7. Binary fractions.

PROBLEMS

1. Print out in triangular form the coefficients of x in the expansion of $(1 + x)^n$ for integral values of n from 1 to 18.

2. Write a program to find the highest common factor and the lowest common multiple of 5 integers. Adapt the program to find the highest common factor and lowest common multiple of n integers, where n is supplied as data.

3. Convert the integers from 1 to 500 into binary form, using
(a) a method of repeated division by 2,
(b) a method of repeated division by 8, followed by conversion from octal into binary digits.

4. Convert the integers between 1 and 500 from base 10 into base b, where b is an integer between 3 and 9 supplied as data. Print the results and check them by converting back to base 10.

5.(a) Investigate the near-Pythagorean triads in the form
$$a^2 + b^2 = c^2 \pm 1$$
where a, b and c are two-digit integers. Print all values of a, b and c and, when a = b, print the value of c/a.

$$\left(\underset{c \to \infty}{\text{Limit}} \; \frac{c}{a} \to \sqrt{2} \right)$$

(b) Find all the Pythagorean triads with elements less than 200. Use the direct method of program 4.3, but find whether the elements possess a common factor by the method of program 4.2.

6. The German mathematician Cantor showed that all rational numbers in the form p/q can be counted. He listed the rationals in diagonal form and numbered them according to the following scheme:

Rationals						Cantor's integers					
1/1	1/2	1/3	1/4	.		1	2	6	7	.	.
2/1	2/2	2/3	.	.		3	5	8	.	.	.
3/1	3/2	.	.	.		4	9	.	.	.	.
4/1	.	.	.	.		10	.	.	.	.	.
.	.	.	.	.		.	.	.	.	.	.

Write a program to print Cantor's integers for all integral values of p and q up to 24.
Hint: The integer (I) corresponding to the rational number $\frac{p}{q}$ is given by the formula

$$I = \sum_{n=1}^{p+q-2} n + (\text{if } p+q \text{ is even then } q, \text{ otherwise } p)$$

7. A large number may be called 'round' if all its prime factors are small. For example, 1200 ($= 2^4.3.5$) would be called round. Find how many round numbers having prime factors less than 10 exist in the range 10^2 to 10^3.

8. 'Amicable' numbers are pairs of numbers such that each is the sum of all the factors of the others. e.g. $220 = 2 \times 2 \times 5 \times 11$ and $284 = 2 \times 2 \times 71$. The factors of 220 are 1, 2, 4, 5, 10, 11, 20, 22, 44, 55, 110, and their sum is 284. The factors of 284 are 1, 2, 4, 71, 142, and their sum is 220. Find all pairs of amicable numbers in the range 100 to 1500.

9. Write a program to find any integers, i, with three digits, a, b and c such that

$$i = a^k + b^k + c^k$$

where k is any integer.

10. A magic square of order n contains n^2 consecutive integers arranged so that the sums of the rows, columns and diagonals are equal. If n is odd and the figure 1 is placed in the middle of the top row, a square may be constructed as shown:

17	24	1	8	15
23	5	7	14	16
4	6	13	20	22
10	12	19	21	3
11	18	25	2	9

Write a program to construct magic squares of order 11 and of order n, where n is odd.

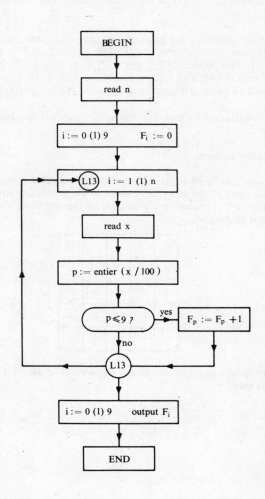

Fig. 5.1(a). Frequency distribution.

5

STATISTICS

The computer is an invaluable aid to the statistician who is thereby
freed from the tedium of the calculations involved in his research.
He is able to use highly sophisticated ideas and formulae, and apply
these in different ways to the data under investigation.

In this chapter, some of the methods used in statistical data pro-
cessing are illustrated, and extensive use is made of arrays for the
intermediate storage of results. All the programs are of an elementary
nature and may be readily adapted to the reader's own problems.

```
begin    comment A program to group a list of n integers in the
         range 0–999 into a frequency table with ten class intervals;
         integer i, n, p, x;
         integer array F[0: 9];
         open (20);
         open (30);
         n := read (20);

         for i := 0 step 1 until 9 do F[i] := 0;
         for i := 1 step 1 until n do
         begin    x := read (20);
                  p := entier (x/100);
                  if p ≤ 9 then F[p] := F[p] + 1;
L13: end;
         for i := 0 step 1 until 9 do output (30, F[i]);
         close (20);
         close (30)
end    →
12;
33; 124; 11; 567; 245; 1789; 385; 294; 994; 174; 856; 12; → →
```

Fig. 5.1(b). Frequency distribution.

5.1. FREQUENCY DISTRIBUTION

In this program, a set of integers in the range 0–999 is sorted into a frequency table having ten equal class intervals. The elements of array F[0: 9] are used to store the number of integers in each of the class intervals 0–99, 100–199, . . . 900–999. The function

$$\text{entier}\left(\frac{x}{100}\right)$$

determines the class interval directly from the input data. Note that the ten subscripts of the array F run from 0–9, allowing for the fact that

$$\text{entier}\left(\frac{33}{100}\right) = 0, \quad \text{entier}\left(\frac{967}{100}\right) = 9, \text{ etc.}$$

The variable n records the number of integers on the data tape, and the summation is stopped when the counter i reaches the value n. The output statement in line 14 prints the frequencies stored in the array F, and these results may be used to illustrate the distribution of the data in the form of a bar chart or histogram.

The inequality in line 12 protects the bounds of the array F in case an integer greater than 999 is printed on the data tape by mistake. For example, the number 1789 would give p a value of 17, but no element F[17] has been declared. The program could be adapted to print elements on the data tape which are outside the range 0–999; such a device is often used as one of the many checks to ensure that data has been correctly typed.

5.2. NUMERICAL ORDER

Sometimes called a 'shunt sort', this method inputs numbers directly into an array A in ascending order of magnitude. The size of the required array is given by the first number, n, on the data tape.

For example, consider the effect of lines 11–16 on the number 93 which forms part of the data given at the bottom of Fig. 5.2(b). After 5 numbers have been sorted, the current values of i and j are

76

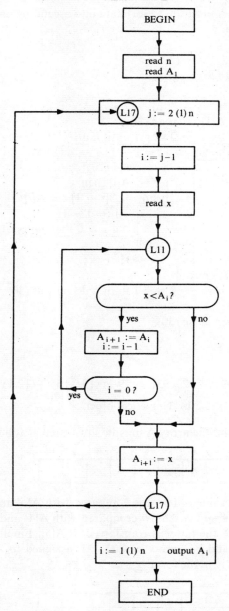

Fig. 5.2(a). Numerical order.

77

```
begin      comment A program to sort data into ascending order;
           integer i, j, n;
           real x;
           open (20);   open (30);
           n := read (20);

           begin      array A[1 : n];
                      A[1] := read (20);
                      for j := 2 step 1 until n do
                      begin      i := j − 1;
                                 x := read (20);
                      L11:       if x < A[i] then
                                 begin      A[i + 1] := A[i];
                                            i := i − 1;
                                            if i ≠ 0 then goto L11
                                 end;
                                 A[i + 1] := x;
                      L17: end;
                      for i := 1 step 1 until n do output (30, A[i])
           end;
           close (20);   close (30)
end →
```

10;
−3.76; 124; 4.27; 99; −12; 93; −79.1; 13; 15; −12.1; → →

Fig. 5.2(b). Numerical order.

5 and 6 respectively, and the array A may be illustrated as follows:

	1	2	3	4	5	6	7	8	9	10
A	−12	−3.76	4.27	99	124					

x is next assigned the value of 93. It is compared with A[5] and as a result 124 is moved to A[6]. x is now compared with A[4] and, since 93 is also less than 99, the latter is transferred to A[5]. Finally, x is compared with A[3], and because x > 4.27, x is assigned to A[4] at line 16, leaving the array in the form

	1	2	3	4	5	6	7	8	9	10
A	−12	−3.76	4.27	93	99	124				

78

5.3. RANKING

This program allots ranks to a set of n numbers already stored in an array in ascending order of magnitude. The value of n is given as the first number on the data tape, and is used to indicate the bounds of the array A. The numbers to be ranked are those shown in the last line of Fig. 5.3(b). The process is similar to that of allotting positions to these numbers and, when two or more equal numbers occur, they share the ranks or positions allotted to them. The first three numbers to be ranked being the same, they share the ranks of 1, 2 and 3, and hence each is given the rank of $\dfrac{1 + 2 + 3}{3} = 2$. The number 13 occurs in positions 9, 10, 11 and 12, and each number 13 is therefore given the rank of $\dfrac{9 + 10 + 11 + 12}{4} = 10\frac{1}{2}$.

The numbers are read into the array A in ascending order of magnitude, and the counter j records either the position in the array of each number, or the position of the first of a set of equal numbers. In the latter case, the counter p records how many numbers are equal to the first of an equal set. The table below shows the current values of i, A[i], j and p when the equal numbers 13 are considered:

i	9	10	11	12
A[i]	13	13	13	13
j	9	9	9	9
p	0	1	2	3

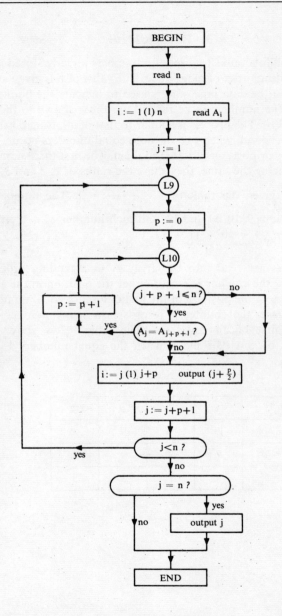

Fig. 5.3(a). Ranking.

The rank for each of the numbers 13 is then found from the expression

$$j + \frac{p}{2}$$

which, in this case, takes the value of $10\frac{1}{2}$. This value is then output for each member of the set of equal numbers.

The clause in line 10

$$\underline{if} \, j + p + 1 \leqslant n \, \underline{and} \, A[j] = A[j + p + 1]$$

prevents the array bounds being exceeded when the possibility of equal numbers is investigated.

```
begin    comment A program to allot ranks to n numbers already in a
         given ordered array;
         integer i, j, n, p;
         open (20);  open (30);
         n := read (20);

         begin   array A[1 : n];
                 for i := 1 step 1 until n do A(i) := read (20);
                 j := 1;
         L9:     p := 0;
         L10:    if j + p + 1 ≤ n and A[j] = A[j + p + 1] then
                 begin   p := p + 1;
                             goto L10
                     end;
                 for i := j step 1 until j + p do output (30, j + p/2);
                 j := j + p + 1;
                 if j < n then goto L9 else if j = n then output (30, j)
         end;
         close (20);  close (30)
end →
```

16;
4; 4; 4; 6.7; 6.7; 7.9; 8; 12.1; 13; 13; 13; 13; 13.7; 14.9; 15; 18; → →

Fig. 5.3(b). Ranking.

At the end of the program, the clause

$$\underline{if} \, j = n \, \underline{then} \, output \, (30, j)$$

allows for the fact that the last number on the data tape will normally be different from its predecessors.

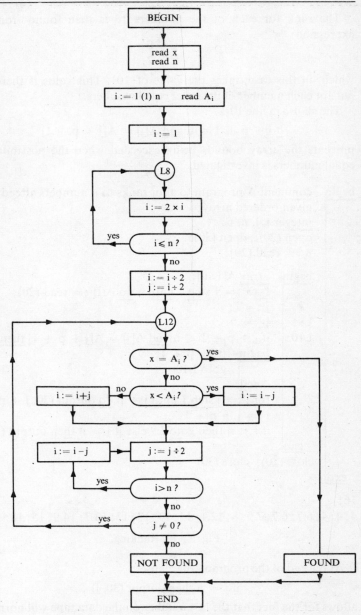

Fig. 5.4. Binary search.

The results of this program are given below and show the ranks allotted to the data printed at the foot of Fig. 5.3(b).

Value	4	4	4	6.7	6.7	7.9	8	12.1
Rank	2	2	2	4.5	4.5	6	7	8

Value	13	13	13	13	13.7	14.9	15	18
Rank	10.5	10.5	10.5	10.5	13	14	15	16

5.4. BINARY SEARCH

This flow diagram finds whether a certain number, x, is present in an array of n numbers stored in ascending order. The most simple method of search would be to compare x with each number in the array in turn, but this process would be time consuming if the required number was not present or was near the end of the array. Program 5.4 uses a method which substantially reduces the search time for both these situations, and is based on the properties of the binary representation of the array subscripts. The method may be illustrated by means of the following simple example in which an array A is filled with 8 numbers in ascending order of magnitude.

Array subscript	Binary representation of subscript	Number stored
1	0001	3
2	0010	8
3	0011	10
4	0100	15
5	0101	16
6	0110	18
7	0111	19
8	1000	21

A search is to be made for the number 19. This number is first compared with the contents of the array element with the largest

83

subscript. This is A[8], and the subscript, 8, can be written as 1000 in binary. The number stored in A[8] is larger than the required number 19, and so the subscript of the array element containing 19 is less than 8. This implies that the binary representation of the required subscript is less than 1000, and is of the form

$$0 \quad p \quad q \quad r$$

where the binary bits p, q and r have yet to be determined. The next element to be tried is A[4]—(4 = 0100 in binary)—and in this case it is found that 19 is larger than the number stored there (15). This implies that p = 1 and that further moves have to be made to determine the values of q and r. It should be noticed that at each stage of the search, a 0 in a particular binary place implies a move to a smaller subscript, whereas a 1 implies a move to a larger subscript. In the search for the number 19 which is stored in A[7]—(7 = 0111 in binary)—q and r are both 1, and so two further moves are made to larger subscripts. The search may be illustrated by the following diagram:

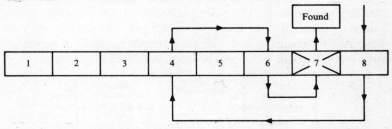

The search will be longest for numbers stored in odd-numbered cells and the number of moves is determined by the number of digits in the binary representation of the subscript excluding final zeros. For example, a search for the number in A[3]—($3_{10} = 0011_2$) would need three moves, but a number in A[8]—($8_{10} = 1000_2$) would be found at once.

The search for the number stored in A[3] is shown in the diagram opposite.

In the program, the number sought, x, and the number of storage cells required, n, are read from the data tape before the array, A, is filled. Counters i and j are used to control the search, and i is initially assigned a power of two next below or equal to n. j is then assigned the value i/2, and is divided by 2 each time a search is made.

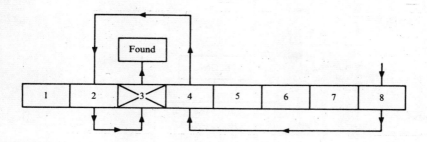

If j becomes zero, the search has not found the given number, and the words 'not found' are printed. The assignments

$$i := i + j; \text{ and } i := i - j;$$

make the move to a larger or smaller subscript, depending on the result of the question

$$x < A_i?$$

The values of i and j in the search for the number stored in A[7] are shown below.

i	j	condition of inequality	resulting assignment
8	4	$x < A[8]$	$i := i - j$
4	2	$x > A[4]$	$i := i + j$
6	1	$x > A[6]$	$i := i + j$
7	0	$x = A[7]$	FOUND

The array bounds are protected by the question

$$i > n?$$

and, if the subscript is outside the bounds, its value is reduced until the search can continue within the given array.

5.5. CORRELATION COEFFICIENT

The flow diagram on page 86 shows a method for calculating the correlation coefficient of a set of ordered pairs of variables, x and y.

85

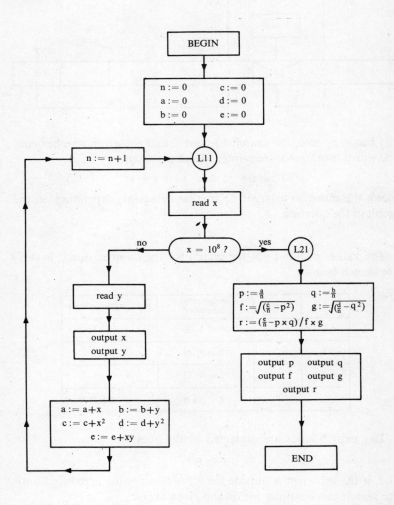

Fig. 5.5. Correlation coefficient.

As a by-product, the mean and standard deviation of each set are found and printed at the end of the calculation. The sums of the x and y values are accumulated directly in the stores a and b, and at the same time, Σx^2, Σy^2 and Σxy are stored in c, d and e respectively. The input of data is terminated if x has the value 10^8, and this number is put on the data tape after the last pair of values of x and y. The terminating value is arbitrary, but in this case 10^8 is chosen because the actual values of x are unlikely to be of this magnitude. The advantage of this method is that it is not necessary to know the number of the pairs of values of x and y on the data tape.

The means, p and q, and the standard deviations, f and g, are calculated using the formulae

$$p = \frac{\Sigma x}{n} \qquad f = \sqrt{\frac{\Sigma x^2}{n} - \left(\frac{\Sigma x}{n}\right)^2}$$

$$q = \frac{\Sigma y}{n} \qquad g = \sqrt{\frac{\Sigma y^2}{n} - \left(\frac{\Sigma y}{n}\right)^2}$$

where n is the number of pairs of values of x and y.

Finally, the correlation coefficient is determined from the formula

$$\frac{\dfrac{\Sigma xy}{n} - pq}{fg}$$

5.6. MARK SCALING

This program is used to scale a set of marks for n candidates in m subjects. The original marks may have any spread and may be out of an arbitrary total.

The first part of the program inputs the data into the even-numbered columns of the array A. At the same time the sum of the marks in each subject is accumulated in array B, and the sum of the squares of these marks in array C. Throughout the program, the candidates are identified by values of the counter i, and the subjects by values of the counter j.

Lines 15–17 determine the mean and standard deviation of the mark distributions in each subject, and these values replace the totals previously stored in arrays B and C. The data tape holds the required

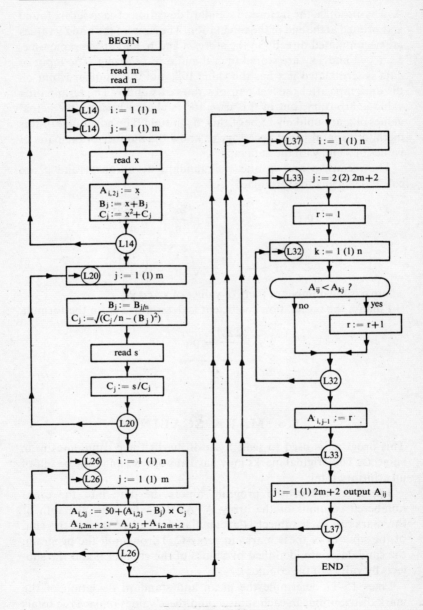

Fig. 5.6(a). Mark scaling.

```
begin    comment To carry out mark scaling for n candidates in m subjects;
         integer i, j, k, m, n, r, s;
         real x;
         open (20);  open (30);
         m := read (20);  n := read (20);
         begin    integer array A[1 : n, 1 : 2 × m + 2], B[1 : m];
                  array C[1 : m];
                  for i := 1 step 1 until n do
                  for j := 1 step 1 until m do
                  begin    x := read (20);
                           A[i, 2 × j] := x;
                           B[j] := x + B[j];
                           C[j] := x↑2 + C[j];
          L14: end        of input;

                  for j := 1 step 1 until m do
                  begin    B[j] := B[j]/n;
                           C[j] := sqrt (C[j]/n − B[j]↑2);
                           s := read (20);
                           C[j] := s/C[j];
          L20: end        of determination of means, standard deviations
                           and scale factors;

                  for i := 1 step 1 until n do
                  for j := 1 step 1 until m do
                  begin    A[i, 2 × j] := 50 + (A[i, 2 × j] − B[j]) × C[j];
                           A[i, 2 × m + 2] := A[i, 2 × j] + A[i, 2 × m + 2];
          L26: end        of scaling and totalling;

                  for i := 1 step 1 until n do
                  begin    for j := 2 step 2 until 2 × m + 2 do
                           begin    r := 1;
                                    for k := 1 step 1 until n do
                                    if A[i, j] < A[k, j] then r := r + 1;
                           L32:     A[i, j − 1] := r;
                           L33: end of ranking;

                           writetext (30, [[cc]]);
                           for j := 1 step 1 until 2 × m + 2 do
                           write (30, format ([nddss]), A[i, j]);
          L37: end        of output
         end;
         close (20);  close (30)
end →
```

Fig. 5.6(b). Mark scaling.

m	n	Latin	French	Physics	Maths	
4;	12;	76;	72;	176;	160;	
		43;	29;	100;	120;	
		103;	57;	96;	77;	
		62;	68;	133;	150;	
		70;	45;	72;	69;	
		25;	40;	60;	70;	
		63;	52;	65;	98;	
		114;	82;	177;	190;	
		137;	82;	165;	188;	
		99;	70;	107;	95;	
		37;	30;	82;	65;	
		133;	55;	130;	145;	
			Standard deviations			
		10;	15;	15;	20;	→ →

Fig. (i). Input for program 5.6.

Latin		French		Physics		Maths		Final Order	
Rank	Mark	Rank	Mark	Rank	Mark	Rank	Mark	Rank	Total
6	44	3	63	2	94	3	69	3	270
10	23	12	25	7	35	6	50	8	133
4	62	6	50	8	32	9	31	7	175
9	35	5	60	4	61	4	64	5	220
7	40	9	39	10	14	11	27	10	120
12	12	10	35	12	5	10	28	12	80
8	36	8	46	11	9	7	41	9	132
3	69	1	72	1	95	1	82	2	318
1	83	1	72	3	85	2	81	1	321
5	59	4	61	6	41	8	39	6	200
11	19	11	26	9	22	12	26	11	93
2	74	7	48	5	58	5	62	4	242

Fig. (ii). Output from program 5.6.

standard deviations in each subject, and these are read in line 18. The necessary scale factors are calculated in line 19, and array C is used again to store them.

The statement

$$A[i, 2 \times j] := 50 + (A[i, 2 \times j] - B[j]) \times C[j];$$

replaces the original mark for the ith candidate in the jth subject by a scaled mark, so that the new distribution of marks has an average of 50 and the required standard deviation. The mark is given to the nearest integer, and the method may be stated as follows:

$$(\text{new mark}) = (\text{required mean}) + (\text{deviation from original mean}) \times (\text{scale factor})$$

where

$$(\text{scale factor}) = \frac{(\text{required standard deviation})}{(\text{original standard deviation})}$$

The last stage of the program determines the candidates' ranks in each subject and also in the final order. The ranking in lines 28–33 is done by counting how many marks are equal; the ranks are then given as one greater than that of the preceding mark. The ranks are stored in the odd-numbered columns of array A (line 32). A typical input and also the corresponding output are printed opposite. The program does not allow for words to be stored on the data tape, or printed in the output, but these have been added for the sake of clarity.

PROBLEMS

1. **'Swop sort'.** There are many methods for sorting data into numerical order, and the following method is given the name of 'swop sort'. The original data are stored in a one-dimensional array A, and may be sorted into descending order by the following operations:
(a) A[1] is compared with A[2] and, if A[1] < A[2], their values are interchanged.
(b) A[2] is now compared with A[3]. If necessary their values are interchanged, and the process is repeated until all adjacent elements have been considered.

(c) Steps (a) and (b) are repeated, and a counter p records the number of interchanges made during each traverse of the array. If, at the end of a traverse, p = 0, no interchanges are necessary and the sorting process is complete.

A simple example is given below:

		1	2	3	4	array subscript
START	A	2	7	9	1	
After 1st loop	A	7	9	2	1	p = 2
After 2nd loop	A	9	7	2	1	p = 1
After 3rd loop	A	9	7	2	1	p = 0 STOP.

The numbers are now sorted in descending order of magnitude.

Write a program to carry out the above procedure on a set of numbers read into an array A.

2. T-tests. T-tests can be used to estimate whether there is a significant statistical difference between two samples from the same parent population. The value of T is given by

$$T = \frac{m_1 - m_2}{\sigma_1 + \sigma_2}$$

where m_1 and m_2 are the means, and σ_1 and σ_2 are the standard deviations of the samples.

It is required to find whether the weights of palæolithic hand-axes from five different sites are significantly different. The numbers in each site are variable, and the weights are given as an integral number of ounces between 1 and 50. Write a program to carry out T-tests between the sites, and to tabulate the results in triangular form. (A value of T = 4 would be thought by an archæologist to show significant difference.)

Site number	1	2	3	4
2		╳	╳	╳
3			╳	╳
4				╳
5				

3. 'Head of the river' boat races. In a 'head of the river' race crews are started on their own at intervals of a few seconds. The time of departure from the start and the time of arrival at the finishing post is noted for each crew, the winner being the boat which covers the course in the shortest time.

Write a program to output the final order of n crews in a head of the river race. Two input tapes should be used: the first holding the crew numbers, c, and their times past the starting post; the second holding the crew numbers together with their times past the finishing post. The data from the first tape should be fed into an array. For each value of c, the starting time should then be replaced by the difference in times, thus filling the array with the true times over the course. As a final stage, the array should be sorted and the order of the crews output with their times over the course.

4. Bi-serial correlation of examination questions. In an examination, certain questions reflect the ability of the candidates to a greater extent than others. The table below shows the results gained on one question of a particular examination, and also the total marks gained on the paper.

Candidate	A	B	C	D	E
Marks gained on question (max 10)	3	9	2	7	6
Total marks (max 100)	45	81	36	38	75

Write a program to correlate the marks gained on each question in an examination with the total marks gained by each candidate using the method of program 5.5. A question which satisfactorily reflects the general ability of the candidate is one which gives a correlation greater than 0.3. In the example given, the correlation coefficient for the particular question is 0.68.

5. Form order by ranking. Write a program to work out a form order for n persons in m subjects using their ranks in each subject instead of their actual marks.

6. Probabilities. Write a program to output the probability of an event occurring r times in n independent trials, when the probability of the event occurring at a single trial is p. Use the binomial distribution formula

$$P_r = {}^nC_r \, p^r \, (1 - p)^{n-r}$$

and generate the probabilities by means of the recurrence relation

$$P_r = P_{r-1} \times \frac{(1 - p)}{p} \times \frac{(n - r + 1)}{r}$$

The values of n and p should be supplied as data, and the probabilities should be printed for all integral values of r between 0 and n.

6

PROCEDURES

INTRODUCTION TO PROCEDURES

Procedures without parameters. A program frequently requires the same series of statements in several different places. To repeat the statements on each occasion would be inconvenient, and so a facility exists for writing them once only at the beginning of the program in the form of a procedure. For example, the arrays A and B may store two matrices whose elements are frequently changing. If the sum of these arrays is needed more than once, a special procedure can be written and given the identifier MATADD.

(*a*) *The procedure body*. The instructions to add the arrays are grouped together within begin ... end brackets, and are known collectively as the procedure body. Suppose the sum of A and B is to be placed in the array C, and that all three arrays have been declared with a bound-pair-list of [1: 3, 1: 4]. The procedure body of MATADD can then be written as

```
begin    integer i, j;
         for i := 1 step 1 until 3 do
         for j := 1 step 1 until 4 do
         C[i, j] := A[i, j] + B[i, j]
end;
```

(*b*) *The procedure declaration*. A procedure identifier, together with its body, is normally declared at the head of a program: alternatively it may be declared at the beginning of the inner block in which it is used. In either case, the identifier and its body are both followed by semi-colons. The order in which declarations are made is unimportant:

```
begin    integer     p, q;
         array       A, B, C[1: 3, 1: 4];
         procedure   MATADD;
                     begin   ... ;
                             ...
                     end;
         real        u, v, w;
```

(c) The procedure call. A procedure is brought into use by means of a procedure call from within the program. Whenever a call occurs, the statements contained in the procedure body are carried out. A call can be made in more than one way but, if no parameters are used, it takes the form of a procedure statement consisting of the identifier alone:

MATADD;

Procedures with parameters. The procedure MATADD has limited use in its present form because it is only capable of adding the arrays A and B. It is clearly advantageous to devise a more general method, and this can be accomplished by means of parameters.

The previous procedure declaration may be adapted so that its identifier is followed in brackets by three arbitrarily chosen parameters, separated by commas:

MATADD (X, Y, Z);

The procedure body must now be written in terms of these same parameters. Thus A, B and C are replaced by X, Y and Z:

```
begin    integer i, j;
         for i := 1 step 1 until 3 do
         for j := 1 step 1 until 4 do
         Z[i, j] := X[i, j] + Y[i, j]
end;
```

The procedure call is likewise re-written. It also contains three parameters, but these are the identifiers of the arrays that are to be added. If the sum of A and B is still to be placed in C, the call is

MATADD (A, B, C);

Similarly, if the sum of arrays D and E is to be placed in F, the call becomes

MATADD (D, E, F);

X, Y and Z are known as formal parameters (FP) and A, B, C (and D, E, F) are called actual parameters (AP). There is always a one-to-one correspondence, both in number and in type, between the formal and the actual parameters:

Formal parameters in procedure declaration	X	Y	Z
	$\updownarrow$	$\updownarrow$	$\updownarrow$
Actual parameters in procedure call	A	B	C
(AP's in alternative procedure call)	(D)	(E)	(F)

Call by value. When using the procedure MATADD, definite values must be assigned to the elements of X and Y before addition can be performed, but no such assignments need to be made to Z. The parameters X and Y may therefore be 'called by value' while Z is 'called by name'. (It is not essential to call X and Y by value as will be seen in program 6.4). The value part of the declaration is written after the procedure identifier and determines the variables that are to be called by value. Variables not mentioned in the value part are called by name:

MATADD (X, Y, Z);

<u>value</u>   X, Y;

Parameters called by value have storage space reserved for them. Thus, when MATADD is called, the values of the elements of A and B are duplicated in the locations occupied by X and Y. A formal parameter omitted from the value part is called by name and each time it occurs in the procedure body it is replaced by the corresponding actual parameter. Thus the identifier C takes the place of Z, and the statements of the procedure body are performed directly on C. In this way the values of A and B are transferred to X and Y, but their sum is placed immediately in C. Z is now a truly formal parameter and is used only to define the action of the procedure body. When the procedure is called, Z has no further part to play.

	Called by value		Called by name
Formal parameters	X	Y	Z
	↑	↑	↕
Actual parameters	A	B	C

Correspondence of parameters. As already stated, there is an exact correspondence between formal and actual parameter lists. If a formal parameter list contains three variables of type <u>real</u>, <u>integer</u> and <u>boolean</u> respectively, the actual parameter list also contains three variables of the same type in the same order. In the case of arrays, the formal and actual parameters must not only be of the same type but must also have similar dimensions.

The procedure MATADD has already been re-written so that it can add any two arrays with bound-pairs of 1: 3, 1: 4. It will now be further adapted to add pairs of arrays of any size.

Suppose that three new arrays, R, S and T, have a bound-pair-list of 1: 6, 1: 5. If the sum of R and S is to be placed in T, additional

96

formal parameters, m and n, are introduced. They take the place of the numbers 3 and 4, and their values are obtained through the corresponding actual parameters of the procedure call.

The procedure identifier is now followed by five formal parameters,

$$\text{MATADD (m, n, X, Y, Z);}$$

the value part becomes

$$\underline{\text{value}}\ \text{m, n, X, Y};$$

and the procedure body is written as

```
begin    integer i, j;
         for i := 1 step 1 until m do
         for j := 1 step 1 until n do
         Z[i, j] := X[i, j] + Y[i, j]
end;
```

Procedure calls contain the five corresponding actual parameters which indicate the arrays to be added and their size:

$$\text{MATADD (3, 4, A, B, C);}$$
$$\text{MATADD (6, 5, R, S, T);}$$

It will be appreciated that, even in its present form, the procedure MATADD is not completely general because both bound-pairs have 1 as the lower bound. Extra parameters may be inserted if different lower bounds are required.

Procedure declarations. The declaration of a procedure containing parameters has been shown to consist of an identifier, a value part (if required) and a procedure body. To these must now be added the specification part which has the effect of declaring the type of the formal parameters. Since there is an exact correspondence between parameter lists, it is unnecessary to include bound-pair-lists after formal parameters specified as arrays. This is because the corresponding actual parameters must be followed by bound-pair-lists when they are declared in the program itself (see Fig. 6.4).

A complete procedure declaration consists of four sections in the following order:

(1) The basic symbol procedure followed by its identifier and formal parameters.

97

(2) The value part.
(3) The specification part.
(4) The procedure body.

The final version of MATADD contains all four sections in its declaration:

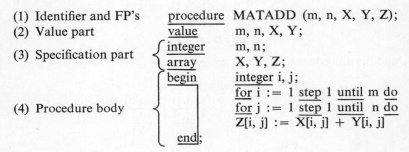

(1) Identifier and FP's   procedure MATADD (m, n, X, Y, Z);
(2) Value part   value   m, n, X, Y;
(3) Specification part   { integer m, n;
                           { array   X, Y, Z;
(4) Procedure body   { begin   integer i, j;
                                      for i := 1 step 1 until m do
                                      for j := 1 step 1 until n do
                                      Z[i, j] := X[i, j] + Y[i, j]
                             end;

Scope of identifiers. When procedures are used, the scope of identifiers can be summarized as follows:

(1) Identifiers declared at the head of a program are universal in scope and hence may be used in a procedure body.
(2) Identifiers declared within a procedure are local to its body and are not accessible outside it.
(3) Specifications are treated as declarations. Thus formal parameters are inaccessible after exit from the procedure body.
(4) The same identifiers may be declared at the head of a program and again, with a different meaning, in a procedure body. An identifier declared in the program is then inaccessible from the procedure body. It retains its value and becomes accessible after exit from the procedure. The identifier declared in the procedure is accessible only within the body.

It should also be noted that, if the same identifier is used as a formal and as an actual parameter, confusion is automatically avoided by a systematic change of identifier within the computer. By this means, a programmer can employ a standard library procedure without fear of mis-using identifiers.

Function designators. In the previous procedure, X and Y were used as input parameters and were called by value; Z was employed as an output parameter and was called by name. In a similar manner, a procedure could be written to find the tangent of an angle, in which case the declaration might be

> <u>procedure</u> TAN (x, v);
> <u>value</u> x;
> <u>real</u> x, v;
> v := sin (x)/cos (x);

A call would take the form of a procedure statement,

> TAN (pi/4, w);

and this would result in w being assigned the value of tan pi/4.

If the result of a procedure is a single number only, an alternative form of procedure may be written to designate a function. In this case, the declaration is preceded by the type of the function, and a value is assigned directly to the procedure identifier:

> <u>real</u> <u>procedure</u> TAN (x);
> <u>value</u> x;
> <u>real</u> x;
> TAN := sin (x)/cos (x);

The procedure statement is replaced by a procedure call known as a function designator, and the identifier, followed by an argument in brackets, is employed to provide a value for direct use in the expression in which it occurs.

> y := TAN (pi/3 + 2);
> output (30, TAN (pi/6));
> z := TAN (y)↑2 − 2 × TAN (y);

The identifier TAN is thus used exactly like a standard function.

It will be noticed that the procedure TAN contains a single statement in its body, and that <u>begin</u> . . . <u>end</u> brackets are therefore omitted. An improved version of this procedure is printed in program 6.1. No flow diagram is provided for this or any program in Chapters 6 or 7 because all are short and relatively simple in structure.

6.1. TAN x

The object of procedure TAN is to permit TAN (x) to be written in a program in place of sin (x)/cos (x). The capital letters have no

```
begin      comment A program to use procedure TAN as a
           function designator and to evaluate the expression
           (tan (a) − tan (b))/(1 + tan (a)×tan (b));

           real procedure TAN (x);
           value x;
           real x;
           if cos (x) = 0
           then begin   writetext (30, [FAIL]);
                        goto L18
                 end
           else TAN := sin (x)/cos (x);

           real a, b, d;
           open (20);  open (30);
           a := read (20);  b := read (20);
           d := 1 + TAN (a) × TAN (b);
           if d = 0 then  writetext (30, [FAIL])
                    else  output (30, (TAN (a) − TAN (b))/d);
L18:       close (20);  close (30)
end        →
```

4.1887902048;   0.2617993878;   → →

Fig. 6.1. Tan x.

significance except that they are used throughout this book in the
identification of procedures, arrays, and labels.

Since TAN identifies a function, it is necessary to declare its type
which, in this case, is real. The program contains two escape clauses
which print failure messages if cos (x) or the divisor, d, is zero.
The program tests the procedure by evaluating

$$\frac{\tan \frac{4\pi}{3} - \tan \frac{\pi}{12}}{1 + \tan \frac{4\pi}{3} . \tan \frac{\pi}{12}}$$

the result of which is known to be $\tan \frac{5\pi}{4}$ $(= -1)$.

100

```
begin      comment A program to test procedure HCF by
           finding the highest common factor of three integers
           x, y and z;
           procedure HCF (a, b) result: (h);
           value a, b;
           integer a, b, h;
           begin
           L8:      if a > b then a := a − b;
                    if a < b then b := b − a;
                    if a = b then h := a else goto L8
               end;

           integer x, y, z, p, q;
           open (20); open (30);
           x := read (20);  y := read (20);  z := read (20);

           HCF (x, y, p);
           HCF (p, z, q);

           output (30, q);
           close (20);  close (30)
end →

15;  30;  95;  →  →
```

Fig. 6.2. Highest common factor.

6.2. HIGHEST COMMON FACTOR

The highest common factor of three numbers was found during the course of program 4.2 in which Pythagorean triads were generated. Euclid's method was used, and this technique is sufficiently general to warrant being turned into a procedure. Program 6.2 contains a procedure to find the highest common factor of two integers only. Two procedure calls are therefore made in order to find the highest common factor of the three integers x, y and z.

The formal parameters a and b are called by value and take

101

initially the values of x and y. Their highest common factor is then assigned to h which, as an output parameter, is called by name.

The first call takes the form of the procedure statement

$$HCF (x, y, p);$$

and assigns to p the highest factor of x and y. The subsequent call

$$HCF (p, z, q);$$

assigns to q the highest factor of p and z.

The procedure makes use of a special Algol facility for explanation. In addition to the comment convention it is possible to insert a series of Algol basic symbols as descriptive matter between formal or actual parameters. A comma is replaced by reversed brackets, between which is placed a sequence of symbols followed by a colon. Thus the procedure declaration may be written as

$$HCF (a, b) result: (h);$$

and the symbols between the reversed brackets will be ignored by the computer.

6.3. FACTORIALS

This program contains two different procedures for calculating factorial n, and shows the difference between the procedure FAC which is used to designate a factorial function, and the procedure FACT which is called by a procedure statement. Both achieve the same results and calculate factorial n in the same way. The chief differences lie in

(1) The inclusion of the type in the declaration of the procedure FAC.
(2) The procedure calls and the methods of transferring the results to the main program.
(3) The output statements.

When procedure FAC is used as a function designator, its identifier, of type integer, has the value of the factorial assigned to it. It is then employed to provide information directly to the program. Alternatively, a call on FACT by means of the statement

$$FACT (n, f);$$

uses a parameter f, through which the result is transferred to the

```
begin    comment A program to find factorial n for
         n := 1 (1) 10, using FAC as a function designator,
         and then again using the procedure FACT;

         integer procedure FAC (n);
         value  n;
         integer n;
         begin    integer f, j;
                  f := 1;
                  for j := 1 step 1 until n do f := f × j;
                  FAC := f
         end;

         procedure  FACT (n, f);
         value  n;
         integer n, f;
         begin    integer j;
                  f := 1;
                  for j := 1 step 1 until n do f := f × j
         end;

         integer n, f;
         open (30);
         for n := 1 step 1 until 10 do output (30, FAC (n));
         writetext (30, [[cc]]);

         for n := 1 step 1 until 10 do
         begin    FACT (n, f);
                  output (30, f)
         end;
         close (30)
end      → →
```

Fig. 6.3. Factorials.

program. Thus the identifier FAC may form part of an output statement but an extra statement is required to print the results from FACT.

It will be noticed that the same identifiers, n, f and j, are used in
the program and in both of the procedures. This is permissible because a procedure is treated as a block, and identifiers used in one
block are inaccessible from blocks in which they are re-declared.

6.4. MATRIX ADDITION

This procedure for the addition of two arrays is a variation on that
described in the introduction to this chapter and differs from it in
three respects:

(a) A label is introduced as a parameter.
(b) All arrays are called by name.
(c) No declarations are made in the procedure body.

(a) The identifier L represents a label and is specified as such in the
procedure declaration. Since the format expression in the program
limits the output to numbers of three or fewer digits, an escape clause
is included in the procedure. A value of 999.5 would be rounded by
the format expression to 1000, and thus a jump is made to L if any
element of Z equals or exceeds this value.

The procedure MATADD is used twice during the program and
on each occasion a jump is made to a different line. The identifier L
is therefore a variable and is included in the formal parameter list.
The corresponding actual parameters are L28 and L34, and the
procedure calls are

$$\text{MATADD (m, n, A, B, C, L28);}$$
and
$$\text{MATADD (m, n, A, A, A, L34);}$$

(b) It is seldom essential to call a parameter by value, but it is
important to do so when the corresponding actual parameter
changes in value between two occurrences of the formal parameter
in the procedure body. To allow for this possibility, it is customary
for input parameters to be called by value. Large arrays may be
exceptions to the rule because they occupy too much storage space
and time is wasted in the copying process. Thus the arrays X, Y and
Z are all called by name in this procedure.

Each procedure call causes the array identifiers in the procedure

```
begin    comment A program to show the use of procedure
         MATADD;
         procedure MATADD (m, n, X, Y, Z, L);
         value   m, n;
         integer m, n;
         array   X, Y, Z;
         label   L;
         for i := 1 step 1 until m do
         for j := 1 step 1 until n do
         begin    Z[i, j] := X[i, j] + Y[i, j];
                  if abs(Z[i, j]) ⩾ 999.5 then goto L.
            end;

         integer f, i, j, m, n;
         open (20);   open (30);
         f := format ([ − ndd]);
         m := read (20);   n := read (20);
         begin    array A, B, C[1: m, 1: n];
                  for i := 1 step 1 until m do
                  for j := 1 step 1 until n do A[i, j] := read (20);
                  for i := 1 step 1 until m do
                  for j := 1 step 1 until n do B[i, j] := read (20);

                  MATADD (m, n, A, B, C, L28);
                  for i := 1 step 1 until m do
                  begin  for j := 1 step 1 until n do
                          write (30, f, C[i, j]);
                          writetext (30, [[c]])
                     end;

         L28:     MATADD (m, n, A, A, A, L34);
                  for i := 1 step 1 until m do
                  begin for j := 1 step 1 until n do
                          write (30, f, A[i, j]);
                          writetext (30, [[c]])
                     end;
         L34: end;
            close (20);   close (30)
end →
```

Fig. 6.4. Matrix addition.

body to be replaced by the corresponding actual parameters. The second call, for instance, has the effect of replacing each of the identifiers X, Y and Z by the identifier A. The statement

$$A[i, j] := A[i, j] + A[i, j];$$

is then executed for all values of i and j, thereby causing the elements of array A to be doubled.

(c) As already stated, identifiers declared at the head of a program are universal in scope unless re-declared elsewhere. The identifiers i and j are included in the initial program declarations and are therefore accessible inside the procedure body without re-declaration.

PROBLEMS

1. Write procedures to designate the following functions: (a) cotangent; (b) reciprocal; (c) cube root; (d) logarithm (base 10).

2. Rewrite programs 3.5 (sin x) and 3.6 (e^x) as real procedures.

3. Write a procedure MATSUB (j, A, B, C) which will add two matrices if j is even, but will otherwise subtract them.

4. Write a procedure to calculate the area of a quadilateral ABCD when the angle DAB and the lengths of all the sides are known.

5. Sub-factorial n, (!n), is defined by the relation
$$!n = n \times !(n - 1) + (-1)^n \text{ where } !0 = 1.$$
Write a procedure SUBFAC and use it to find !7 and !12.

6. An aircraft leaves a point O and steers x° E of N for t minutes with an airspeed of u knots. If the wind velocity is v knots from a direction of y° E of N, write a procedure to find the distance travelled and the bearing of the aircraft from O.

7

ADVANCED PROCEDURES

7.1. GRAPH PLOTTING

Many large computers have a special graph plotting device available. Others, not so well equipped, can only type the results using standard symbols, and for these the plotting of a graph is an exercise in printing.

Program 7.1 contains a graph plotting procedure. The function to be plotted is declared as one of the actual parameters, and the procedure calculates the ordinates for abscissae at unit intervals between given limits, a and b.

The procedure uses the formal parameters

$$x, y, a, b,$$

and the corresponding actual parameters of the program are

$$t, f(t), -40, 40,$$

where $f(t) = 15e^{-t/90} \cos t/5$.

The graph of $y = f(t)$ is to be plotted, and it is necessary to include t as one of the actual parameters. The limits, a and b, are called by value, but the formal parameters x and y are called by name so that they may be replaced by t and f(t). Since x (and hence t) changes in value during the procedure, f(t) will do the same, and so a different value will be assigned to y each time it occurs. This method by which one parameter is made to depend upon another is known as Jensen's device.

The array A may be thought of as being superimposed upon graph paper. Each element of A is initially assigned a value of 1, and elements retaining this value are later represented as empty spaces. Elements along the x-axis are given a value of 2 (to be printed later with a minus sign) and those along the y-axis a value of 3 (printed as a colon). The origin contains a 4 (a plus sign) and each point on the graph is represented in the array and printed as a 0.

In order to save both time and storage space, the ordinates of the

```
begin    comment A procedure to plot the graph of y = f(x) for
         a ≤ x ≤ b;
         procedure GRAPHPLOT (x, y, a, b);
         value    a, b;
         real     x, y, a, b;
         begin    integer array A[−24: 24, a: b];
                  integer m, n;
                  for m := a step 1 until b do
                  for n := −24 step 1 until 24 do A[n, m] := 1;

                  for m := a step 1 until b do A[0, m] := 2;
                  for n := −24 step 1 until 24 do A[n, 0] := 3;
                  A[0, 0] := 4;
                  for x := a step 1 until b do
                  if abs(y) ≤ 24 then A[y, x] := 0;

                  for n := 24 step −1 until −24 do
                  begin    writetext (30, [[c]]);
                           for m := a step 1 until b do
                           if A[n, m] = 1 then writetext (30, [ * ]) else
                           if A[n, m] = 2 then writetext (30, [−]) else
                           if A[n, m] = 3 then writetext (30, [ : ]) else
                           if A[n, m] = 4 then writetext (30, [+]) else
                           if A[n, m] = 0 then writetext (30, [ 0 ])
                  end
         end;

         real t;
         open (30);
         GRAPHPLOT (t, 15 × exp(−t/90) × cos(t/5), −40, 40);
         close (30)
end      → →
```

Fig. 7.1. Graph plotting.

Graph of $y = 15e^{-t/90} \cos t/5$.

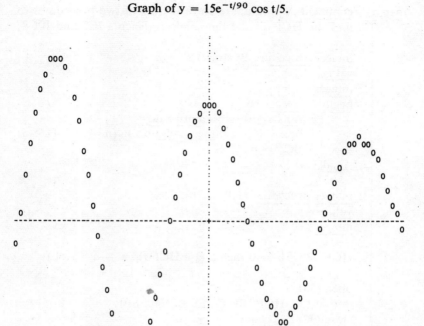

graph are limited to values between ± 24. Also, the multi-conditional clause at the end of the procedure is written bearing in mind that, once a condition is satisfied, later conditions are not investigated. Thus, when printing the graph, spaces are considered first because they occur most frequently.

The results cannot of course be accurate because only integer values are plotted but, as the diagram shows, the important features of a function can be well illustrated.

```
begin    comment A program to show the use of two procedures to
         find the HCF of the binomial coefficients 7C2 and 10C8;

         integer procedure BC (n, r);
         value    n, r;
         integer  n, r;
         begin    integer bc, j;
                  for j := 0 step 1 until r do
                  bc := if j = 0 then 1 else bc × (n − j + 1) ÷ j;
                  BC := bc
            end;

         integer procedure HCF (a, b);
         value    a, b;
         integer  a, b;

         HCF := if b = 0 then a else HCF (b, a − a ÷ b × b);

         open (30);
         output (30, HCF (BC (7, 2), BC (10, 8)));
         close (30)
end      → →
```

Fig. 7.2. Recursive highest common factor.

7.2. RECURSIVE HIGHEST COMMON FACTOR

This program provides another method for finding the highest common factor of two numbers and makes use of a device by which a procedure may be called recursively from within its own body:

$$HCF := \underline{if}\ b = 0\ \underline{then}\ a\ \underline{else}\ HCF\ (b, a - a \div b \times b);$$

The action of this statement is not easy to follow because of the repeated use of the same identifiers, but its effect may be illustrated by considering the action of the call HCF (10, 15).

```
HCF (10, 15) := ..................... HCF (15, 10)
          := ..................... HCF (10,  5)
          := ..................... HCF ( 5,  0)
          := if b = 0 then 5 else ...............
```

110

The program also contains an integer procedure for generating a binomial coefficient. This is not recursive because no fresh call on the procedure is made from within its body. Care must nevertheless be taken to use a temporary identifier, bc, to hold the coefficient as it is calculated. The value of bc is not assigned to the procedure identifier BC until the former has been fully evaluated.

Both the procedures are employed as function designators, and the program calculates and prints the highest common factor of $^7C_2$ and $^{10}C_8$ by means of the single statement,

output (30, HCF (BC (7, 2), BC (10, 8)));

7.3. MONTE CARLO

Monte Carlo methods use, as the name suggests, a chance element in the solution of a problem. The success of such methods depends upon access to a set of random numbers which may be generated when required or taken from a printed table.

Random methods are frequently used in the field of nuclear physics where the behaviour of fundamental particles may be simulated using mathematical models based on the laws of probability. Likewise, random numbers are generated electronically by

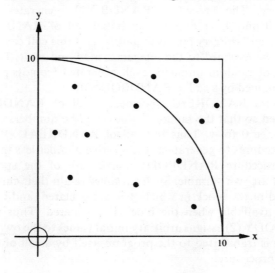

111

Ernie, the machine owned by the British Post Office which picks the winners of prizes in monthly Premium Bond draws.

Program 7.3 illustrates the use of random numbers in the calculation of π. The figure shows a square enclosing a quadrant of a circle of radius 10 units. Each point (x, y) is represented by a pair of random co-ordinates between 0 and 10. Two counters are used: p records the number of points generated by a random number procedure, and q records the number of these points which lie on or within the boundary of the circle. If enough points are taken

$$\frac{p}{q} \doteq \frac{10^2}{\frac{1}{4}\pi\, 10^2}$$

and an approximation to π is then given by the formula

$$\pi \doteq \frac{4q}{p}$$

This program considers 3000 points. It prints an answer after every 20 points and finally finds π to an accuracy of about 1 per cent.

The procedures for the generation of random numbers are re-printed from the tape library of the Oxford University Computing Laboratory. The procedure RANDBODY generates a number between 0 and 1. It requires an initial call of RANDBODY (v) where v is an arbitrary positive number, and this call causes random values to be assigned to the integers z, mod and mult. Thereafter, a sequence of random numbers, each generated from its predecessor, can be obtained by a call of RANDBODY (0).

Procedure RANDREAL includes a call of RANDBODY (0) and is used so that the random values shall be numbers of type _real_ in the range 0 to w. Thus the call of RANDREAL (10) employs both procedures to generate a real positive number less than 10.

The procedure RANDBODY makes use of the special Algol facility of an _own_ variable. Such variables retain their current values at the end of the block in which they are declared, and these values are then available when the block is re-entered. Thus the call of RANDBODY (2) assigns arbitrary initial values to the _own_ variables. No value is transferred to the program itself by a call of this nature on a real procedure.

112

```
begin    comment A program to find an approximate value for pi
         using a Monte Carlo method;

         real procedure RANDBODY (v);
         integer v;
         begin    own integer z, mult, mod;
                  integer i;
                  if v > 0 then
                  begin    z := 1; mult := 3↑7; mod := 2↑20;
                           for i := v step −1 until 2 do
                           begin    z := z × mult + 1;
                                    z := z − z ÷ mod × mod
                           end
                  end    of initial entry;
                  z := z × mult + 1;
                  z := z − z ÷ mod × mod;
                  RANDBODY := (z + .5)/mod
         end;

         real procedure RANDREAL (w);
         real w;
         comment In range (0, w);
         RANDREAL := RANDBODY (0) × w;

         integer p, q;
         real x, y;
         q := 0;
         RANDBODY (2);
         open (30);

         for p := 1 step 1 until 3000 do
         begin    x := RANDREAL (10);
                  y := RANDREAL (10);
                  if x↑2 + y↑2 ≤ 100 then q := q + 1;
                  if p ÷ 20 × 20 − p = 0 then output (30, 4 × q/p)
         end;
         close (30)
end      → →
```

Fig. 7.3. Monte Carlo.

```
begin    comment A program to test a table for associativity;
         procedure ASSOC (A, x, y, z, b);
         value A, x, y, z;
         integer array A;
         integer x, y, z;
         boolean b;
         b := A[A[x, y], z] = A[x, A[y, z]];

         integer n, x, y, z, f;
         boolean b;
         f := format ([snd]);
         open (20);   open (30);
         n := read (20);
         begin    integer array A[1: n, 1: n];
                  for x := 1 step 1 until n do
                  for y := 1 step 1 until n do
                  A[x, y] := read (20);

                  for x := 1 step 1 until n do
                  for y := 1 step 1 until n do
                  for z := 1 step 1 until n do
                  begin    ASSOC (A, x, y, z, b);
                           if not b then
                           begin    writetext (30, [FAIL]);
                                    write (30, f, x);
                                    write (30, f, y);
                                    write (30, f, z);
                                    goto L29
                                    end
                           end;
         L29:     write boolean (30, b);
                  close (20);   close (30)
                  end
end    →
```

Fig. 7.4. Associativity.

It should be noted that the numbers generated by the procedure RANDBODY are properly described as pseudo-random numbers. This is because, once an arbitrary value has been assigned to the parameter v, a fixed sequence of numbers is generated. It is impossible to generate truly random numbers on a general-purpose computer.

7.4. ASSOCIATIVITY

A set of elements $\{x, y, z \ldots\}$ form a group under the action of a binary operation, $*$, if all the following properties are true:

(a) The set is closed under the operation $*$.
(b) The operation is associative.
(c) The set possesses a neutral element.
(d) Each element has a unique inverse.

The associative property is generally tedious to test because, in a group table of order n (i.e. containing n different elements), the number of possible relationships of the form

$$x * (y * z) = (x * y) * z$$

is n^3. Program 7.4 employs the procedure ASSOC to carry out tests for this property and, though not essential, boolean variables are included so that their use may be illustrated.

The identifier b is declared as type boolean and takes one of the logical values true or false depending upon the truth of the statement

$$b := A[A[x, y], z] = A[x, A[y, z]];$$

All combinations of elements are tested and, if the equality is false, (i.e. if not b), the corresponding elements are printed. The truth of the table is finally output as the logical value of b, using the instruction

write boolean (30, b);

Like all other input and output instructions, this statement makes use of a special KDF 9 procedure, and takes the place of the statement

if b then writetext (30, [true])
else writetext (30, [false]);

The table to be tested for associativity is expressed in numerical form and, when punched as data, is preceded by its order. The neutral element is placed in the top left-hand corner of the table and is

115

allotted the number 1. The other elements are then numbered as they occur along the top line (or down the left-hand column).

Table

Operation *		Second element e a b c d
First element	e	e a b c d
	a	a e c d b
	b	b c d e a
	c	c d a b e
	d	d b e a c

Data

5;

1; 2; 3; 4; 5;
2; 1; 4; 5; 3;
3; 4; 5; 1; 2;
4; 5; 2; 3; 1;
5; 3; 1; 2; 4;

In the example above, e is the neutral element and the results show that the table is not associative. The first case of failure found by the program is when

$$x = 2, \qquad y = 2, \qquad z = 3$$

i.e.

$$b * (b * c) \neq (b * b) * c$$

7.5. SIMPSON

Simpson's rule states that for an even number, n,

$$\int_a^b f(x)\,dx = \frac{1}{3}\frac{b-a}{n}\left[(y_0 + y_n) + 4(y_1 + y_3 + \ldots + y_{n-1}) + 2(y_2 + y_4 + \ldots + y_{n-2})\right]$$

The real procedure SIMPSON uses this rule to evaluate an integral and it doubles the value of n until two successive approximations differ by less than a given small quantity, d. The function to be integrated and its derivative must be finite for $a \leqslant x \leqslant b$. The parameter list contains the function, y, and its argument, x, as well as a, b and d. Jensen's device is used (see program 7.1), and x and y are both called by name.

116

```
begin comment A program to test procedure SIMPSON which
      integrates the function y between the limits x = a and
      x = b. The number of ordinates is doubled until two
      approximations differ by less than a given small quantity, d;

      real procedure SIMPSON (y, x, a, b, d);
      value a, b, d;
      real   y, x, a, b, d;
      begin integer j, n;
            real h, s, area, previous area;
            area := 0;
            for n := 2, 2 × n while abs (area − previous area) > d do
            begin array A[0: n];
                  h := (b − a)/n;
                  previous area := area;
                  for j := 0 step 1 until n do
                  begin   x := a + j × h;
                          A[j] := y
                  end;
                  s := A[0] + A[n];
                  for j := 1 step 1 until n − 1 do
                  s := s + A[j] × (if j ÷ 2 × 2 − j = 0 then 2 else 4);
                  area := s × h/3
            end;
            SIMPSON := area
      end;

      real x;
      open (30);
      output (30, SIMPSON (4/(1 + x↑2), x, 0, 1, 10−6));
      close (30)
end → →
```

Fig. 7.5. Simpson.

117

The counter, j, determines the successive values of x, and the corresponding ordinates are stored in A[j]. Ordinates, other than the first and last, are multiplied by 2 if j is even, and by 4 if j is odd. The sum of the ordinates is then placed in s. The value of the integral is finally assigned to the identifier SIMPSON which is used as a function designator in the program.

The program employs the procedure SIMPSON to calculate π to six decimal places by finding

$$\int_0^1 \frac{4}{1 + x^2}\, dx.$$

7.6. MULTI-LENGTH ADDITION

The normal working accuracy of a computer is limited by the number of binary digits that can be held in each cell. It is, however, always possible to increase the accuracy by means of arrays in which numbers are stored digit by digit.

In this program, procedure ADD is used to find the sum of 1/15 + 4/9 to 20 decimal places. The two fractions, expressed as decimals, are fed, digit by digit, into arrays X and Y, and their sum is placed in Z by the statement

$$\text{ADD } (10, -20, X, Y, Z);$$

The initial parameters signify that base 10 is being used for the calculation and that terms down to 10^{-20} are to be included. It is not essential to restrict the working to base 10. The procedure is equally effective with other bases, and is particularly useful when the base is a power of 10.

Conversion into base 2 was illustrated in program 4.6 and, once a decimal number has been converted into binary form, it is a simple matter to change it into bases 4, 8, 16, . . . , 2^n. This is done by dividing the binary digits into groups of 2, 3, 4, . . . , n, and writing down the decimal equivalents of each group. For example:

$$
\begin{aligned}
2685_{10} &= |1|0|1|0|0|1|1|1|1|1|1|0|1| = 101001111101_2 \\
&= |1\ 0|1\ 0|0\ 1|1\ 1|1\ 1|0\ 1| = 221331_4 \\
&= |1\ 0\ 1|0\ 0\ 1|1\ 1\ 1|1\ 0\ 1| = 5175_8 \\
&= |1\ 0\ 1\ 0|0\ 1\ 1\ 1|1\ 1\ 0\ 1| = 10, 7, 13_{16}
\end{aligned}
$$

```
begin    comment A program to test the multi-length procedure
         ADD by evaluating 1/15 + 4/9 to 20 decimal places;

         procedure ADD (b, n, A, B, C);
         value    b, n, A, B;
         integer  b, n;
         integer  array A, B, C;
         begin    integer i;
                  for i := n step 1 until −1 do
                  begin    C[i] := A[i] + B[i];
                           if C[i] ⩾ b then
                           begin    B[i + 1] := B[i + 1] + 1;
                                    C[i] := C[i] − b
                           end
                  end
         end;

         integer array X, Y, Z[−20 : −1];
         integer k;
         for k := −20 step 1 until −1 do
         begin    X[k] := 6;
                  Y[k] := 4
         end;
         X[−1] := 0;  X[−20] := 7;

         ADD (10, −20, X, Y, Z);

         open (30);
         writetext (30, [0.]);
         for k := −1 step −1 until −20 do
         write (30, format ([d]), Z[k]);
         close (30)
end      → →
```

Fig. 7.6. Multi-length addition.

119

In the same way, decimal numbers can be converted into base 100 or 1000.

$$|2|1|7|9|4|2|3| = 2179423_{10}$$
$$|.2|1\ 7|9\ 4|2\ 3| = 2, 17, 94, 23_{100}$$
$$|..2|1\ 7\ 9|4\ 2\ 3| = 2, 179, 423_{1000}$$

By this means, a 7-digit number in base 10 becomes a 4-digit number in base 100 and a 3-digit number in base 1000. In the procedure, digits are separated by means of arrays, and the maximum size of the base is governed only by the limitations of the computer.

Suppose procedure ADD employs a base b of 10^5. 20 decimal places are now equivalent to 4 places in base b and the procedure call becomes

$$\text{ADD } (_{10}5, -4, X, Y, Z);$$

This signifies that only 4 array elements are needed and permits the arrays X, Y and Z to be declared with a bound-pair of $-4: -1$. Decimal digits are stored 5 at a time in X and Y, and their sum is output from Z in the same way. Both the time and the number of storage locations are reduced by nearly four-fifths, and the advantages of a large base become apparent.

7.7. MULTI-LENGTH DIVISION

The final procedure in this chapter is to divide an integer into a real number stored by digits in an array. It is used (together with procedure ADD which is not reprinted in full) to find an approximation to the first 100 digits of e.

The dividend is placed in array A, and the procedure divides each element, A[j], by the divisor, d. It places each quotient, q, in B[j] and records the remainder, $A[j] - q \times d$. The product of the remainder and the base, b, is then added to the next element of the dividend, $A[j - 1]$. The new value of $A[j - 1]$ can never exceed $b + d \times b$ and is the largest number that can occur within the procedure. This number must be less than the maximum number that can be stored in each cell and therefore determines the largest base that can be used.

```
begin    comment A program to find an approximation to the first
         100 digits of e using procedure DIV and base ₁₀9;

         procedure ADD (b, n, A, B, C);

         procedure DIV (b, n, d, A, B);
         value    b, n, d, A;
         integer  b, n, d;
         integer  array A, B;
         begin    integer j, m, q;
                  m := n + 1;
                  for j := −1 step −1 until m do
                  begin    q := A[j] ÷ d;
                           A[j−1] := A[j−1] + b×(A[j] − q×d);
                           B[j] := q
                  end;
                  B[n] := A[n]/d
         end;

         integer i, k;
         integer array X, Y[−11: −1];
         for i := −1 step −1 until −11 do
         X[i] := Y[i] := 666666666;
         Y[−1] := 166666666;
         open (30);
         for k := 4 step 1 until 70 do
         begin    DIV (₁₀9, −11, k, Y, Y);
                  ADD (₁₀9, −11, X, Y, X);
                  if k ÷ 10×10 − k = 0 then
                  begin    writetext (30, [[cc]2.]);
                           for i := −1 step −1 until −11 do
                           write (30, format ([sddddddddd]), X[i])
                  end
         end;
         close (30)
end      → →
```

Fig. 7.7. Multi-length division.

121

A feature of the program is the use of the array Y to store first the dividend and then the quotient. Thus the procedure call

$$\text{DIV (b, n, d, Y, Y);}$$

results in the array Y being reduced by a factor of d. The working in the procedure is in base b, an accuracy of b^n is required and all arrays have a bound-pair of n: -1. In practice, b is 10^9, n is -11 and 99 decimal places are obtained. The result turns out to be correct to this number of places, though extra terms would normally be taken to ensure accuracy.

The sum of the first four terms of the exponential series

$$1 + \frac{1}{1!} + \frac{1}{2!} + \frac{1}{3!} \cdots$$

is 2.6 recurring. The figure 2 is conveniently ignored in the ensuing calculations, but is printed in the final answer by the statement

$$\text{writetext (30, [[cc]2.]);}$$

The array X is used to hold the current sum of the terms, and initially the decimal fraction 0.6 recurring is stored in it nine digits at a time. $1/3!$ is stored in array Y in the same way. It is divided by 4 and then by each successive integer in turn until the desired accuracy has been obtained. Since $1/70! < 10^{-99}$, the counter k is allowed to run from 4 to 70. Finally, after each division, the next term of the series is added to the cumulative total in X. This total is printed after every 10 terms in order to illustrate the convergence of the series.

Calculation of e

Number of terms added	10	20	30	40	50	60	70
Number of significant figures obtained	8	20	34	50	66	83	100

PROBLEMS

1. If a set of elements is known to form a group under the binary operation, $*$, it is also said to be Abelian if the commutative property

$$x * y = y * x$$

is satisfied by all combinations of elements. Write a procedure to test a group table for this property.

2. Use a random integer generator to find the approximate probability that a number less than 10^6 will exceed the cube of the sum of its digits.

3. Adapt the procedure GRAPHPLOT
  (a) to superimpose two functions on the same axes;
  (b) to plot a circle or ellipse.

4. (a) Write a recursive procedure to evaluate factorial n.
  (b) Use the principles of multi-length arithmetic to calculate the first 100 factorials in floating point form to 10 significant figures.
  (c) Calculate accurately the first 200 factorials.
  (d) Repeat (a), (b) and (c) for sub-factorials.

5. Write a procedure CONICPLOT (D) to plot a conic of the general form $ax^2 + by^2 + 2hxy + 2gx + 2fy + c = 0$, whose constants a, b, h, g, f, and c are stored in the array D.

6. Write a program in which six random numbers are generated, stored in an array D, and plotted by the procedure CONICPLOT. Restrictions may be placed on the range of the random numbers in order that the resulting conic can be plotted without the use of scale factors.

7. Write procedures to find the area under a curve using
  (a) the trapezium rule

$$A = \frac{h}{2} (y_0 + 2y_1 + y_2)$$

  (b) the rule for seven ordinates

$$A = \frac{h}{140} (41(y_0 + y_6) + 216(y_1 + y_5) + 27(y_2 + y_4) + 272y_3)$$

where h is the width of a strip. The number of strips should be successively doubled in (a) and increased by 6 in (b).

8. Write multi-length procedures:
  (a) to multiply an integer of m digits by an integer less than 100,
  (b) to multiply any two integers,
  (c) to multiply any two real numbers.

9. Write multi-length procedures:
  (a) to find the reciprocal of an integer with m digits to n decimal places,
  (b) to divide any two real numbers.

10. Use the Gregory-Machin series,

$$\frac{\pi}{4} = 4 \tan^{-1}\frac{1}{5} - \tan^{-1}\frac{1}{239}$$

to calculate π to 30 decimal places.

123

8

ITERATION

Iteration techniques are of great importance in numerical work, and some of them are illustrated in this chapter. The method of false position is used to develop one iterative formula, and another is derived from Newton's approximation to the root of an equation. The general iterative formula

$$x_{r+1} = F(x_r)$$

is discussed, and an outline is given of the convergence of such methods.

THE METHOD OF FALSE POSITION

Consider the problem of finding the value of $\sqrt{5}$ by solving the equation $x^2 - 5 = 0$. This is equivalent to finding the co-ordinate of the intersection of the curve $y = x^2 - 5$ with the positive x-axis (Fig. (i)).

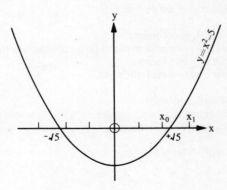

Fig. (i).

The method is as follows:

(1) Find a pair of bounds within which a close approximation to the required root lies, say $x_0 = 2$ and $x_1 = 3$.

(2) Find the values of the ordinates, y_0 and y_1 at $x = x_0$ and $x = x_1$. In this case, $y_0 = -1$ and $y_1 = 4$.

(3) Let P_0 and P_1 be the points (x_0, y_0) and (x_1, y_1) respectively, and call P_0 the 'pivot'.

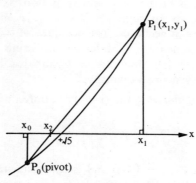

Fig. (ii).

(4) The line $P_0 P_1$ will meet the x-axis in a point $(x_2, 0)$, where x_2 is given by the formula

$$x_2 = \frac{x_0 y_1 - x_1 y_0}{y_1 - y_0} \qquad \text{(Fig. (ii)).}$$

(This may be verified by direct proportion.)

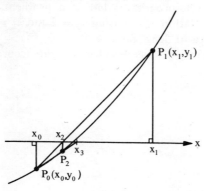

Fig. (iii).

125

ITERATION

(5) The ordinate at $x = x_2$ is now calculated, and the line P_0P_2 is drawn to meet the x-axis in x_3, where

$$x_3 = \frac{x_0y_2 - x_2y_0}{y_2 - y_0} \qquad \text{(Fig. (iii))}.$$

(6) P_3 is now found in a similar manner, and P_0P_3 is drawn to intersect the axis at the next approximation, x_4. This process is repeated until an approximation $x = x_n$ is obtained with the required degree of accuracy. The accuracy of the approximation may be determined by considering the value of the function $x_n^2 - 5$, and the iteration can be terminated if $|x_n^2 - 5| < d$ where d is a given small quantity.

The formula for deriving the $(r + 1)^{\text{th}}$ approximation from x_r is given by

$$x_{r+1} = \frac{x_0y_r - x_ry_0}{y_r - y_0}$$

This is of the form $x_{r+1} = F(x_r)$ and is called an iteration formula. This formula is used repeatedly until the required accuracy is achieved.

Care must be taken to ensure that the formula used is convergent, i.e. after a finite number of iterations, the next approximation is an improvement on the previous value. In the simple case above, it can be shown that the process necessarily converges to the correct root, because only one root of the quadratic equation lies between the chosen values of x_0 and x_1.

The method of false position is simple, but not very fast, and the writing of a suitable program is left as a problem. The successive results in the calculation of $\sqrt{5}$ using $x_0 = 2$ and $x_1 = 3$ are tabulated opposite (working to 5 sig. fig.).

After 4 iterations, $\sqrt{5}$ has been found to an accuracy of 4 decimal places.

r	x_r	y_r
0	2.0000	−1.0000
1	3.0000	4.0000
2	2.2000	−0.1600
3	2.2381	0.0091
4	2.2360	−0.0003
5	2.2361	0.0001

THE GENERAL ITERATION FORMULA

When finding roots of an equation by means of an iterative formula of the form

$$x_{r+1} = F(x_r),$$

it may be necessary to rearrange the equation in several different ways until a convergent sequence of approximations to a root is found. Consider the equation

$$x^3 + x - 1 = 0.$$

A sketch of the function $y = x^3 + x - 1$ shows that this equation has a real root between $x = 0$ and $x = 1$. The root could be found by the method of false position, using the point $P_0(1,1)$ as pivot, and $x_1 = 0.5$ as a first estimate (see Fig. (iv)).

However, the equation $x^3 + x - 1 = 0$ may be rearranged in the form

$$x = \frac{1}{1 + x^2}$$

or

$$x = (1 - x)^{\frac{1}{3}}$$

or

$$x = 1 - x^3$$

and these suggest the iteration formulae

$$x_{r+1} = \frac{1}{1 + x_r^2} \qquad (1)$$

$$x_{r+1} = (1 - x_r)^{\frac{1}{3}} \qquad (2)$$

and

$$x_{r+1} = 1 - x_r^3. \qquad (3)$$

127

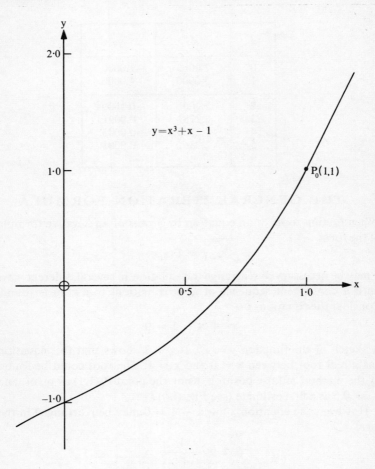

Fig. (iv).

The successive steps in the solution may be illustrated by means of a diagram showing the path of the iteration. The solution of the equation

$$x = \frac{1}{1 + x^2}$$

may be represented by the intersection of the line $y_a = x$ and the curve $y_b = 1/(1 + x^2)$ (see Fig. (v)).

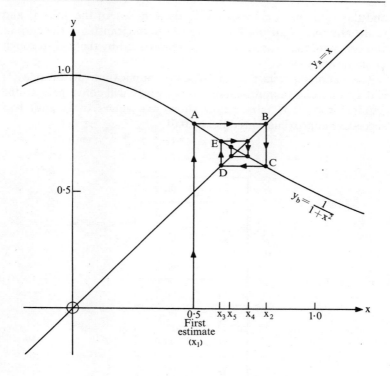

Fig. (v).

A first estimate of the value of the root may be taken as $x_1 = 0.5$. Formula (1) then gives the next estimate as

$$x_2 = y_b(x_1) = \frac{1}{1 + (0.5)^2} = 0.8.$$

This value corresponds to the ordinate of the point A (i.e. the current value of the function y_b), and also to the abscissa of the point C. The transfer of the current value of y_b to the next estimate of the root may be represented in the figure by the directed lines $\overline{AB}$ and $\overline{BC}$, where B is the point on the line $y_a = x$ having the same ordinate as the point A and the same abscissa as the point C. A further application of the iteration formula gives

$$x_3 = y_b(x_2) = \frac{1}{1 + (0.8)^2} = 0.61 \quad \text{(2 sig. fig.)}$$

and this value of x_3 corresponds to the ordinate of the point C and to the abscissa of the point E. Here again, the transfer of the current value of y_b to the next value of x_r is represented by the path through D on the line $y_a = x$.

The figure suggests that the repeated application of formula (1) will give a series of approximations, x_r, which will converge onto the required root. The table below shows the values of x_r until two successive approximations differ by 0.01.

x_1	0.50
x_2	0.80
x_3	0.61
x_4	0.73
x_5	0.65
x_6	0.70
x_7	0.67
x_8	0.69
x_9	0.68

The convergence of the process depends on the slope of the curve $y_b = 1/(1 + x^2)$ in the neighbourhood of the root. In the interval from $x = 0$ to $x = 1$, the slope lies between 0 and -1 and this fact is sufficient to ensure convergence onto a root in this interval from a reasonably close first estimate.

In general, it can be shown that the iteration formula

$$x_{r+1} = F(x_r)$$

is convergent if $|F'(x)| < 1$ in the neighbourhood of a required root, and a close enough first estimate is made. Figs. (vi) and (vii) show the path of the iterations given by formulae (2) and (3). Formula

130

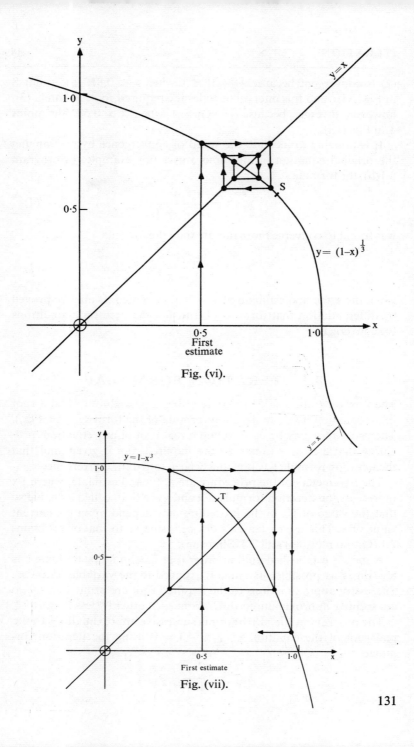

y

1·0

0·5

y = x

S

y = (1-x)^{\frac{1}{3}}

0·5
First
estimate

1·0

x

Fig. (vi).

y

1·0

0·5

y = 1-x^3

y = x

T

0·5
First estimate

1·0

x

Fig. (vii).

131

(2) is convergent because $|F'(x)| < 1$ when $x < 0.81$, (the point S in Fig. (vi)), and this interval includes the required root. Formula (3), however, diverges because $|F'(x)| > 1$ when $x > 0.58$ (the point T in Fig. (vii)).

It is possible to increase the speed of convergence by rearranging the original equation in ingenious ways. For example, in program 8.1(b), the formula

$$x_{r+1} = \frac{x_r^3 + 1}{2x_r^2 + 1} \tag{4}$$

was found to converge more rapidly than the formula

$$x_{r+1} = \frac{1}{(1 + x_r^2)} \tag{1}$$

when the same first estimate of $x = 0.5$ was taken. It may be proved by differentiation that formula (4) satisfies the necessary conditions for convergence.

8.1. ITERATION FORMULAE

The flow diagram in Fig. 8.1(a) describes a procedure to find a root of an equation using an iteration formula of the form $x_{r+1} = F(x_r)$. The procedure may be used to find a real root of any equation, providing that a close estimate to the required root is given, and that the iteration formula has previously been tested for convergence.

The parameters of the procedure are x, Fx and estimate, where Fx represents the iteration formula. Fx and x are both called by name so that the value of Fx in the procedure may depend upon the current value of x. This use of Jensen's device is similar to that of programs 7.1 (Graph plotting) and 7.5 (Simpson).

A loop is established and, in order that Fx should be evaluated as few times as possible, its value is assigned to the variable 'latest x'. Successive approximations to the required root are printed until two consecutive approximations differ from each other by less than 10^{-8}.

The program in Fig. 8.1(b) prints successive approximations to the real root of the equation $x^3 + x - 1 = 0$ using the iteration formulae

$$x_{r+1} = \frac{1}{1 + x_r^2} \tag{1}$$

and
$$x_{r+1} = \frac{x_r^3 + 1}{2x_r^2 + 1} \qquad (4)$$

In each case, an initial estimate of $x = 0.5$ is used, and the results show that the numbers of iterations needed to achieve the required accuracy are 39 and 14 respectively.

A summary of the results is given below:

No. of iterations	Iteration formula (1)				Iteration formula (4)			
1	+7.9999	9999	999	$_{10}-1;$	+7.5000	0000	000	$_{10}-1;$
2	+6.0975	6097	561	$_{10}-1;$	+6.6911	7647	059	$_{10}-1;$
3	+7.2896	7909	799	$_{10}-1;$	+6.8563	4184	360	$_{10}-1;$
4	+6.5299	9724	809	$_{10}-1;$	+6.8153	8261	872	$_{10}-1;$
5	+7.0106	1372	973	$_{10}-1;$	+6.8251	8584	032	$_{10}-1;$
6	+6.7047	1795	842	$_{10}-1;$	+6.8228	1834	670	$_{10}-1;$
7	+6.8987	7632	249	$_{10}-1;$	+6.8233	8887	811	$_{10}-1;$
8	+6.7753	8380	913	$_{10}-1;$	+6.8232	5131	722	$_{10}-1;$
9	+6.8537	3592	716	$_{10}-1;$	+6.8232	8448	042	$_{10}-1;$
10	+6.8039	3856	950	$_{10}-1;$	+6.8232	7648	519	$_{10}-1;$
11	+6.8355	6995	191	$_{10}-1;$	+6.8232	7841	270	$_{10}-1;$
12	+6.8154	7034	987	$_{10}-1;$	+6.8232	7794	801	$_{10}-1;$
13	+6.8282	3937	608	$_{10}-1;$	+6.8232	7806	004	$_{10}-1;$
14	+6.8201	2619	072	$_{10}-1;$	+6.8232	7803	303	$_{10}-1;$
15	+6.8252	8067	204	$_{10}-1;$				
---	---	---	---	---				
37	+6.8232	7813	112	$_{10}-1;$				
38	+6.8232	7797	929	$_{10}-1;$				
39	+6.8232	7807	575	$_{10}-1;$				

THE NEWTON–RAPHSON ITERATION FORMULA

Let α be a root of the equation $f(x) = 0$, and let x_r be an approximation to the root such that $|\alpha - x_r|$ is small.
Then
$$f(\alpha) = f(x_r + \overline{\alpha - x_r}) = 0.$$

By Taylor's theorem,
$$f(x_r + \overline{\alpha - x_r}) = f(x_r) + (\alpha - x_r)f'(x_r) + \frac{(\alpha - x_r)^2}{2!} f''(x_r) + \dots \qquad (5)$$

133

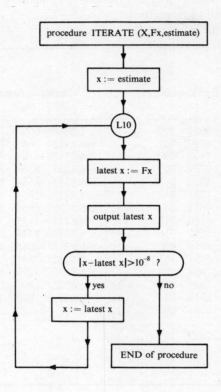

Fig. 8.1(a). Iteration formulae.

```
begin    comment A program to test a procedure for finding a root
         of an equation. The procedure requires an estimate of the
         root and a convergent iteration formula in the form
         x = F(x);
         procedure ITERATE (x, Fx, estimate);
         value    estimate;
         real     x, Fx, estimate;
         begin    real latest x;
                  x := estimate;
         L10:     latest x := Fx;
                  output (30, latest x);
                  if abs(x − latest x) > 10−8 then
                  begin    x := latest x;
                              goto L10
                        end
            end;

         real x;
         open (30);
         ITERATE (x, 1/(1 + x↑2), 0.5);
         writetext (30, [[cc]]);
         ITERATE (x, (x↑3 + 1)/(2 × x↑2 + 1), 0.5);
         close (30)
end        → →
```

Fig. 8.1(b). Iteration formulae.

and so, to a first approximation,

$$\alpha = x_r - \frac{f(x_r)}{f'(x_r)} \tag{6}$$

This equation suggests the formula

$$x_{r+1} = x_r - \frac{f(x_r)}{f'(x_r)} \tag{7}$$

which is known as the Newton–Raphson iteration formula.

Consider the solution of the equation $x^2 - a = 0$.

$$f(x) = x^2 - a$$
$$f'(x) = 2x$$

135

If x_r is a first approximation, the next approximation is given by

$$x_{r+1} = x_r - \frac{(x_r^2 - a)}{2x_r}$$

$$= \frac{1}{2}\left(x_r + \frac{a}{x_r}\right)$$

This is the iteration formula used in program 8.2—Square Roots.

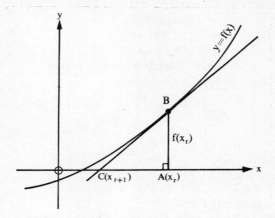

Fig. (viii).

Graphical Interpretation. Fig. (viii) shows the relationship between a given approximation x_r and the next value x_{r+1} given by the iteration formula

$$x_{r+1} = x_r - \frac{f(x_r)}{f'(x_r)}.$$

If A is the point $(x_r, 0)$, $AB = f(x_r)$ and $\tan \angle BCA = f'(x_r)$. Hence

$$CA = \frac{AB}{\tan \angle BCA}$$

But $$OC = OA - CA$$

Thus
$$OC = x_r - \frac{f(x_r)}{f'(x_r)}$$

and so the x co-ordinate of C corresponds to the next approximation x_{r+1}.

Fig. (ix) shows how the repeated application of the iteration formula gives a series of approximations that converge onto the required root. Fig. (x) illustrates how an approximation near one root can converge onto a neighbouring root. In the latter case, a very close approximation would be required to ensure convergence onto the required root.

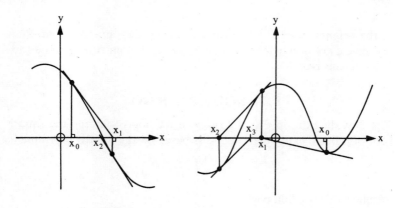

Fig. (ix).   Fig. (x).

The convergence of the Newton–Raphson formula. It should be noticed that equation (7) may be written in the form

$$(\alpha - x_{r+1}) = (\alpha - x_r) + \frac{f(x_r)}{f'(x_r)} \qquad (8)$$

and that equation (5) may be written as

$$(\alpha - x_r) = -\frac{f(x_r)}{f'(x_r)} - \frac{(\alpha - x_r)^2}{2!}\frac{f''(x_r)}{f'(x_r)} - \ldots \text{ higher terms } (9)$$

137

Equations (8) and (9) now give

$$(\alpha - x_{r+1}) = -\frac{(\alpha - x_r)^2}{2!} \frac{f''(x_r)}{f'(x_r)} - \dots \text{ higher terms,}$$

and show that if the approximation x_r is near to the root, the error at the next stage in the iteration is proportional to the square of the error at the previous stage. This property is known as quadratic convergence, and makes the Newton–Raphson formula a very powerful iteration method.

It can be shown that the condition for convergence of the Newton–Raphson formula is that

$$|f(x).f''(x)| < [f'(x)]^2$$

in the neighbourhood of the root. In most cases, graphical intuition will give a good indication of convergence, though such methods are by no means rigorous.

8.2. SQUARE ROOT

The flow diagram in Fig. 8.2 shows a method of finding the square root of a positive number, a, and is based on the iterative formula

$$x_{r+1} = \frac{1}{2}\left(x_r + \frac{a}{x_r}\right)$$

which was derived on page 136.

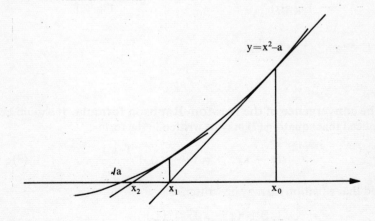

Fig. (xi).

It can be seen from a sketch of the function $y = x^2 - a$ that the Newton–Raphson process gives a sequence of decreasing approximations $x_0, x_1, x_2, \ldots$ which converge onto $\sqrt{a}$ from above if the first approximation, x_0, is taken to be greater than $\sqrt{a}$. This is ensured by putting

$$x_0 = \tfrac{1}{2}(1 + a)$$

because $\tfrac{1}{2}(1 + a) \geqslant \sqrt{a}$ for all positive values of a. This inequality may be proved as follows:

$$(\sqrt{a} - 1)^2 \geqslant 0$$
$$\Leftrightarrow \quad a - 2\sqrt{a} + 1 \geqslant 0$$
$$\Leftrightarrow \quad a + 1 \geqslant 2\sqrt{a}$$
$$\Leftrightarrow \quad \tfrac{1}{2}(a + 1) \geqslant \sqrt{a}$$

(Equality occurring when $a = 1$).

The value of the number a is read from the data tape, and if a is negative or zero, a fail notice is printed and the program concluded. The identifier latest x is again used to hold the latest estimate of the square root. The iteration converges onto the root from above, and so, if latest x is found to be greater than or equal to x, noise-level accuracy has been reached.

The following table shows how the iterative formula

$$x_{r+1} = \frac{1}{2}\left(x_r + \frac{5}{x_r}\right)$$

converges onto $\sqrt{5}$ from an initial estimate of

$$x_0 = \tfrac{1}{2}(1 + 5) = 3.$$

r	x_r
0	3.0000
1	2.3333
2	2.2381
3	2.2361

These results may be compared with those obtained from the method of false position (see page 127).

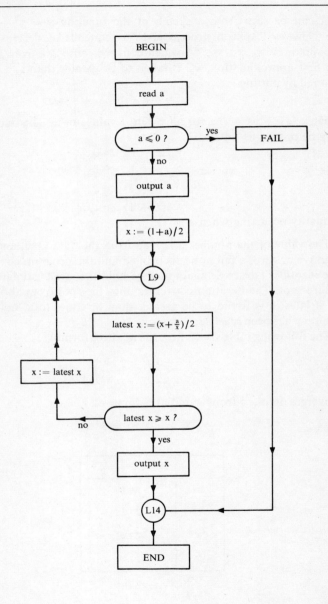

Fig. 8.2. Square root.

8.3. NEWTON–RAPHSON

Program 8.3 uses the iteration formula

$$x_{r+1} = x_r - \frac{f(x_r)}{f'(x_r)}$$

to find roots of functions, given close estimates. The parameter list of the procedure contains the function, its derivative, the argument of the function and the estimate. Each parameter except the estimate must be called by name so that values may be assigned to them after entry into the procedure body (Jensen's device).

The procedure NEWTON is used in this program to output directly the zeros of the three functions

$$f(x) = \sin x - x - 2$$
$$f(t) = t^3 - \cos t$$

and

$$f(z) = \tan \frac{z}{4} - 1.$$

Provision is made in line 12 to jump out of the iterative loop when an accuracy of 10^{-8} has been achieved. If convergence is too slow, or has not yet begun, the conditions in the following line stop the iteration after 20 loops.

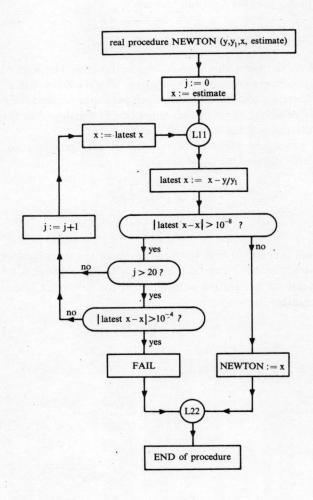

Fig. 8.3(a). Newton–Raphson.

```
begin  real procedure NEWTON (y, y1, x, estimate);
       value   estimate;
       real    y, y1, x, estimate;
       comment A procedure to find a root of a continuous
       single-valued function using the Newton–Raphson iteration
       formula and given a close estimate;
       begin   integer j;
               real latest x;
               j := 0;
               x := estimate;
       L11:    latest x := x − y/y1;
               if abs (latest x − x) > ₁₀−8 then
               begin   if j > 20 and abs (latest x − x) > ₁₀−4
                       then begin   writetext (30, [FAIL]);
                                    goto L22
                            end;
                       j := j + 1;
                       x := latest x;
                       goto L11
               end;
               NEWTON := x;
       L22: end

       real t, x, z;
       open (30);
       output (30, NEWTON (sin (x) − x − 2, cos (x) − 1, x, − 2.5));
       output (30, NEWTON (t↑3 − cos (t), 3 × t↑2 + sin (t), t, 1));
       output (30, NEWTON (sin (z/4)/cos (z/4)−1, 0.25/(cos (z/4))↑2,
       close (30)
end    → →
```

Fig. 8.3(b). Newton–Raphson.

PROBLEMS

1. Write a program to find the square root of a positive number, n, by the method of false position. Use $x = 0$ as the pivotal value, and $x = n$ as a first estimate. This will ensure convergence to the required positive root. Continue the iteration until the root is given to 5 significant figures, and print out the values of each successive approximation.

2. Use the method of false position to find the real root of the equation

$$y = x^3 + x - 1$$

between $x = 0$ and $x = 1$, using the point $(1,1)$ as pivot, and $x = 0.5$ as a first estimate.

3. Find the real root of the equation

$$x - e^{-x} = 0$$

by the following methods, comparing the speed of convergence in each case.
(a) Use the method of false position taking $x = 1$ as pivotal value, and $x = 0$ as a first estimate.
(b) Use the Newton–Raphson method to solve the equation taking $x = 1$ as a first estimate.
(c) Use the iteration formula
$$x_{r+1} = e^{-x_r}$$
and put $x_1 = 0$.
Draw graphs of the functions, and in (c), show that the necessary conditions of convergence are satisfied.

4. Repeat question 3 to find the root of the equation

$$x - \cos x = 0$$

between $x = 0$ and $x = \pi/2$.
(a) For the method of false position, take $x = 1$ as pivotal value, and and $x = 0.5$ as a first estimate.
(b) For the Newton–Raphson method, take $x = 0.5$ as a first estimate.
(c) Use the iteration formula
$$x_{r+1} = \cos x_r$$
and put $x_1 = 0.5$.

5. (a) Write a program based on flow diagram 8.2 to find the square roots of n numbers supplied as data.
(b) Adapt the program to find fourth roots.
(c) Adapt the program further to find r^{th} roots, where r is a power of 2 supplied as data.

9

POLYNOMIALS

Polynomial equations of degree less than five can be solved by direct methods, and the first part of this chapter shows how formulae can be used to solve quadratic and cubic equations. The rest of the chapter is concerned with the solution of the general polynomial, and a program is given to show how all the real roots are found using the Newton–Raphson iteration method. The extension to complex roots is left as a problem, but a procedure for polynomial division is given in detail and is used as a basis for several other problems.

9.1. QUADRATIC

This program, given in flow diagram form only, solves a quadratic equation by means of the well-known formula

$$x = [-b \pm \sqrt{(b^2 - 4ac)}]/2a$$

The different types of roots which can occur are determined by the relative values of a, b and d. If both a and b are zero, the word 'fail' is printed. When the discriminant, d ($= b^2 - 4ac$), is negative, the program calculates the real and imaginary parts of the resulting complex roots.

No attempt is made at error control, and there is a possibility of loss of significant figures in roots close to zero. This will result from the expression

$$x = (-b + \sqrt{d})/2a \text{ when } b \doteq \sqrt{d} \text{ and } b > 0$$

or $\qquad x = (-b - \sqrt{d})/2a \text{ when } b \doteq \sqrt{d} \text{ and } b < 0.$

The difficulty can be avoided by finding the numerically larger root first and calculating the smaller root from the fact that the product of the roots is c/a. The writing of a suitable program is left as a problem.

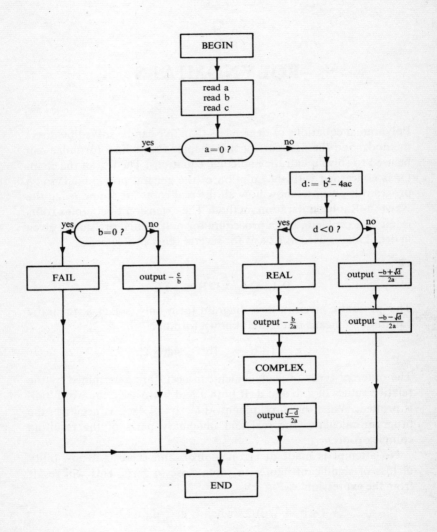

Fig. 9.1. Quadratic equation.

The cubic equation. The cubic equation

$$ax^3 + 3bx^2 + 3cx + d = 0 \quad (a \neq 0) \qquad (1)$$

may be solved by first reducing the coefficient of x^2 to zero using the substitution

$$x = (y - b)/a \qquad (2)$$

The resulting form of the equation is

$$y^3 + 3hy + g = 0 \qquad (3)$$

where $\qquad\qquad h = ac - b^2$

and $\qquad\qquad g = a^2d - 3abc + 2b^3.$

By putting $\qquad\qquad y = p + q \qquad (4)$

and using the identity

$$(p + q)^3 = p^3 + q^3 + 3pq(p + q)$$

equation (3) reduces to

$$y^3 - 3pqy - (p^3 + q^3) = 0 \qquad (5)$$

Comparing the coefficients in equations (3) and (5),

$$p^3 + q^3 = -g \qquad (6)$$
$$pq = -h \qquad (7)$$
or $\qquad\qquad p^3 q^3 = -h^3 \qquad (8)$

From equations (6) and (8) it follows that p^3 and q^3 are roots of the quadratic equation

$$z^2 + gz - h^3 = 0 \qquad (9)$$

The roots, z_1 and z_2, of equation (9) may be found from the equations

$$z_1 = \tfrac{1}{2}[-g + \sqrt{(g^2 + 4h^3)}] \quad \text{and} \quad z_2 = -g - z_1.$$

The symmetry of equations (6) and (8) shows that either z_1 or z_2 may be taken as the value of p^3. If $z_1 = p^3$, the corresponding value of q^3 is $-g - p^3$, and p and q are given by the cube roots of z_1 and z_2.

If $g^2 + 4h^3 > 0$, then z_1 and z_2 are real and distinct. The corresponding real values of p and q are then $p = (z_1)^{\frac{1}{3}}$ and $q = (z_2)^{\frac{1}{3}}$ and the complex values are given by

$$\omega p, \ \omega^2 p \quad \text{and} \quad \omega^2 q, \ \omega q$$

where ω and ω^2 are the complex cube roots of unity. Since

$$y = p + q \tag{4}$$

the roots of equation (3) are given by

$$
\begin{aligned}
y_1 &= p + q \\
y_2 &= \omega p + \omega^2 q \\
y_3 &= \omega^2 p + \omega q
\end{aligned}
$$

The root ωp corresponds to $\omega^2 q$ because equation (7) must be satisfied by all pairs of roots of equation (9), i.e.

$$pq = (\omega p)(\omega^2 q) = (\omega^2 p)(\omega q) = -h.$$

It can be shown that

$$\omega = \frac{1}{2} + j \frac{\sqrt{3}}{2}$$

and $\qquad \omega^2 = \dfrac{1}{2} - j \dfrac{\sqrt{3}}{2} \qquad$ where $j = \sqrt{-1}$.

Since $\qquad\qquad\qquad x = (y - b)/a \tag{2}$

the corresponding roots of equation (1) then reduce to

$$x_1 = [(p + q) - b]/a \tag{10}$$

$$x_2 = -[\tfrac{1}{2}(p + q) + b]/a + j\sqrt{[3}(p - q)]/2a \tag{11}$$

$$x_3 = -[\tfrac{1}{2}(p + q) + b]/a - j\sqrt{[3}(p - q)]/2a \tag{12}$$

Thus if $g^2 + 4h^3 > 0$, equation (1) has one real and two imaginary roots.

If $g^2 + 4h^3 \leqslant 0$, the roots of the equation (1) are all real, but the above method gives the roots in complex form, and is unsuitable for computational purposes. In this case, a solution can be found more easily as follows.

Solution by trigonometrical substitution when $g^2 + 4h^3 \leqslant 0$.
Putting $y = 2(-h)^{\frac{1}{2}} \cos \theta$, equation (3) becomes

$$8(-h)^{\frac{3}{2}} \cos^3 \theta - b(-h)^{\frac{1}{2}} \cos \theta = -g.$$

After reduction, this becomes

$$2(-h)^{\frac{3}{2}} \cos 3\theta = -g$$

and finally
$$\cos 3\theta = \frac{-g}{2(h)^{\frac{3}{2}}}. \tag{13}$$

Since $g^2 \leqslant -4h^3$ it follows that $\left| \dfrac{-g}{2(-h)^{\frac{3}{2}}} \right| \leqslant 1$

and so a real value of θ can be found to satisfy equation (13). If ϕ is one such value, the roots of equation (3) are given by

$$y_1 = k \cos \phi$$

$$y_2 = k \cos (\phi + \tfrac{2}{3}\pi) \quad \text{where } k = 2\sqrt{(-h)}$$

$$y_3 = k \cos (\phi + \tfrac{4}{3}\pi)$$

The corresponding values for x are then found from equation (2). The value of ϕ is found from the equation

$$3\phi = \text{arc cos} \left[\frac{-g}{2(-h)^{\frac{3}{2}}} \right] \tag{14}$$

but the only inverse function in the set of Algol standard functions is arc tan. It is therefore necessary to make the following conversion.

Let
$$c = \frac{-g}{2(-h)^{\frac{3}{2}}} \quad \text{where } c = \cos 3\theta.$$

Then
$$\tan 3\phi = \frac{(1 - c^2)^{\frac{1}{2}}}{c}$$

and
$$\phi = \tfrac{1}{3} \text{ arc tan } [(1 - c^2)^{\frac{1}{2}}/c]$$

149

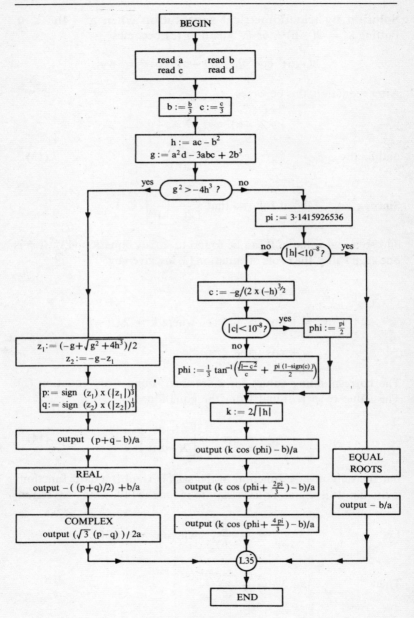

BEGIN

read a   read b
read c   read d

$b := \frac{b}{3}$ $c := \frac{c}{3}$

$h := ac - b^2$
$g := a^2 d - 3abc + 2b^3$

$g^2 > -4h^3$?   yes   no

pi := 3·1415926536

$|h| < 10^{-8}$?   no   yes

$c := -g/(2 \times (-h)^{3/2})$

$|c| < 10^{-8}$?   yes   phi := $\frac{pi}{2}$
no

$phi := \frac{1}{3} \tan^{-1}\left(\frac{\sqrt{1-c^2}}{c} + \frac{pi\,(1-sign(c))}{2}\right)$

$z_1 := (-g + \sqrt{g^2 + 4h^3})/2$
$z_2 := -g - z_1$

$p := sign\ (z_1) \times (|z_1|)^{\frac{1}{3}}$
$q := sign\ (z_2) \times (|z_2|)^{\frac{1}{3}}$

$k := 2\sqrt{|h|}$

output $(p+q-b)/a$

output (k cos (phi) − b)/a

EQUAL
ROOTS

REAL
output $-(\,(p+q)/2\,) + b/a$

output (k cos (phi + $\frac{2pi}{3}$) − b)/a

output − b/a

COMPLEX
output $(\sqrt{3}\ (p-q)\ )/2a$

output (k cos (phi + $\frac{4pi}{3}$) − b)/a

L35

END

Fig. 9.2(a). Cubic.

```
begin    comment A program to find the solutions of the cubic
         equation a×x↑3 + 3×b×x↑2 + 3×c×x + d = 0   when a ≠ 0;
         real a, b, c, d, g, h, k, p, q, z1, z2, phi, pi;
         open (20); open (30);
         a := read (20); b := read (20); c := read (20); d := read (20);
         b := b/3; c := c/3;

         comment the cubic is first reduced to
         the form y↑3 + 3×h×y + g = 0;
         h := a×c − b↑2;
         g := a↑2×d − 3×a×b×c + 2×b↑3;
         if g↑2 > −4 × h↑3
         then begin  z1 := (−g + sqrt (g↑2 + 4×h↑3))/2;
                     z2 := −g − z1;
                     p := sign (z1) × abs (z1)↑(1/3);
                     q := sign (z2) × abs (z2)↑(1/3);
                     output (30, (p + q − b)/a);
                     writetext (30, [REAL]);
                     output (30, −((p + q)/2 + b)/a);
                     writetext (30, [COMPLEX]);
                     output (30, (sqrt (3) × (p − q))/(2 × a))
               end
         else begin  pi := 3.1415926536;
                     if abs (h) < ₁₀−8 then
                     begin   writetext (30, [EQUAL ROOTS]);
                             output (30, −b/a);
                             goto L35
                     end;
                     c := −g/(2 × (−h)↑(3/2));
                     if abs (c) < ₁₀−8 then phi := pi/2 else
                     phi := (arc tan (sqrt (1−c↑2)/c)+pi × (1−sign(c))/2)/3;
                     k := 2 × sqrt (abs (h));
                     output (30, (k × cos (phi) − b)/a);
                     output (30, (k × cos (phi + (2×pi)/3) − b)/a);
                     output (30, (k × cos (phi + (4×pi)/3) − b)/a);
         L35: end;
         close (20);  close (30)
end   →
```

Fig. 9.2(b). Cubic.

A difficulty arises because the principal values of the functions arc tan and arc cos lie in different ranges. The range for arc tan is $-\pi/2$ to $\pi/2$, whereas the range for arc cos is 0 to π. In the figure, these ranges are shown by heavy lines.

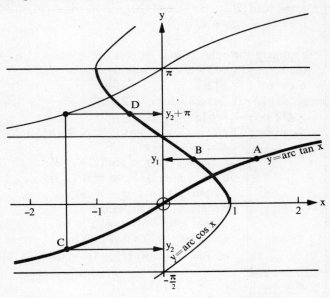

A value y_1 of arc tan in the range $0 \leqslant y \leqslant \pi/2$ corresponds exactly to a principal value y_1 of arc cos x (points A and B in the figure). However, a value y_2 of arc tan x in the range $-\pi/2 \leqslant y \leqslant 0$ corresponds to a principal value $y_2 + \pi$ of arc cos x (points C and D in the figure). Thus any value of ϕ between $-\pi/2$ and 0 resulting from equation (14) must be increased by π.

9.2. CUBIC

The first part of the program deals with the case of one real and two complex roots, and the second part uses a trigonometrical solution for the case of three real roots.

The data tape holds the coefficients of the terms of the cubic equation and these are assigned to a, b, c and d. The values of b and c are then reduced by a factor of 3 to correspond with the values of b and c in equation (1) of the introduction.

When only one root is real, this is printed by the statement

$$\text{output } (30, (p + q - b)/a);$$

The real part of the pair of complex roots is printed by the statement

$$\text{output } (30, -((p + q)/2 + b)/a);$$

and the imaginary part by

$$\text{output } (30, (\text{sqrt } (3) \times (p - q))/(2 \times a))$$

The identifier c is used for two different purposes in this program. After the first part of the program, c is no longer required to store the value of a coefficient, and so may be used later to store the value of $\cos 3\theta$.

Use is made of the function sign (x) which can take only the values $+1$, 0 or -1. This function is first used to find $(z_1)^{\frac{1}{3}}$ using the statement

$$p := \text{sign } (z_1) \times \text{abs } (z_1)\uparrow(1/3);$$

If z_1 is negative, the negative cube root is found, and the difficulty of raising a negative number to a fractional exponent is avoided. The function sign (x) is used again in line 30 to allow for the difference in the principal values of arc cos and arc tan as explained on page 152. The expression

$$\text{pi} \times (1 - \text{sign } (c))/2$$

takes the value π if c is negative, and the value 0 if c is positive.

If $|h| < 10^{-8}$ and $g^2 \leqslant 4h^3$, then g is necessarily small, and the equation in y is of the form

$$y^3 \doteq 0.$$

In this case, the roots are assumed to be equal, and the corresponding values, $x = -(b/a)$, are output in line 25.

Iterative methods for root finding. Providing that complex roots are not required, the solution of a well-behaved polynomial equation is not a major task. On the other hand, the more general a problem becomes, the harder it is to write a perfect program and there will usually be special circumstances that may defeat it.

The Newton–Raphson iteration formula

$$x_{r+1} = x_r - \frac{f(x_r)}{f'(x_r)}$$

may be used to find a real root of a given polynomial equation, and this root may then be removed from the polynomial by division. The reduced equation may have other real roots, and these can be found by further applications of the above equation followed by division. It is not necessary to find a close approximation to any particular root, because the order in which the roots are extracted is unimportant unless a high degree of accuracy is required.

(1) *Polynomial division—linear factor only.* A very simple procedure is involved when a polynomial is to be divided by a linear factor. This may be illustrated by the following example:

Divide $3x^3 + 2x^2 - x - 2$ by $x - 2$.

The full working may be set out as follows:

$$
\begin{array}{r}
3x^2 + 8x + 15 \\
x - 2 \overline{\smash{)}\,3x^3 + 2x^2 - x - 2} \\
\underline{3x^3 - 6x^2} \\
8x^2 - x \\
\underline{8x^2 - 16x} \\
15x - 2 \\
\underline{15x - 30} \\
28
\end{array}
$$

However, the working may be simplified by the use of detached coefficients (sometimes called synthetic division):

Division by the linear factor $x - 2$.

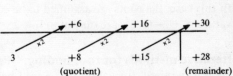

3	+2	−1	−2
	+6	+16	+30
3	+8	+15	+28
	(quotient)		(remainder)

Start with the first coefficient, 3, and multiply it by 2. Put the result, 6, under the next coefficient, 2, and add. Multiply the result, 8, by 2 and add to the next coefficient, -1. Finally, multiply the result, 15, by 2 and add to the last coefficient, -2, to give the remainder, 28.

Consider how a program can be written for the general case when a polynomial of degree n,

$$a_n x^n + a_{n-1} x^{n-1} + \ldots + a_1 x + a_0,$$

is divided by a linear factor $x - \alpha$. The coefficients of the polynomial may be stored in an array A[0: n] such that A[i] holds the value of a_i. A second array B[0: n] stores the coefficients of the quotient (of degree $n - 1$)

$$b_{n-1} x^{n-1} + b_{n-2} x^{n-2} + \ldots + b_1 x + b_0$$

and the remainder r.

Array A	a_n	a_{n-1}	a_{n-2}	----	a_2	a_1	a_0	
	n	n–1	n–2		2	1	0	array subscripts
Array B	b_{n-1}	b_{n-2}	b_{n-3}	----	b_1	b_0	r	

It will be seen from the example on synthetic division that the coefficients in array B may be obtained from those in array A as follows:

$$b_{n-1} = a_n$$
$$b_{n-2} = b_{n-1} \times \alpha + a_{n-1}$$
$$b_{n-3} = b_{n-2} \times \alpha + a_{n-2}$$
$$\ldots\ldots\ldots\ldots\ldots\ldots\ldots\ldots$$
$$b_0 = b_1 \times \alpha + a_1$$

In terms of subscripted variables, the above equations may be written

$$B[n] = A[n]$$
$$B[i] = B[i + 1] \times \alpha + A[i + 1] \quad (n > i \geqslant 0)$$

The coefficients of the quotient are then given by the values of

$$B[n], B[n - 1], \ldots, B[1]$$

and the remainder by the value of B[0].

(2) *Calculation of $f(\alpha)/f'(\alpha)$ by division.* Let α be an approximate root of the polynomial equation

$$f(x) = 0.$$

Then, by division $f(x) = (x - \alpha).q_1(x) + r_1$ (1)

where $q_1(x)$ and r_1 are the quotient and remainder respectively.

Thus $f(\alpha) = r_1$ (2)

Differentiating (1) with respect to x:

$$f'(x) = q_1(x) + (x - \alpha).q'_1(x)$$

Thus $f'(\alpha) = q_1(\alpha)$

Dividing $q_1(x)$ by $(x - \alpha)$:

$$q_1(x) = (x - \alpha).q_2(x) + r_2$$ (3)

where $q_2(x)$ and r_2 are the quotient and remainder.

Thus, from (3), $q_1(\alpha) = r_2$ (4)

From (2) and (4), $\dfrac{f(\alpha)}{f'(\alpha)} = \dfrac{r_1}{r_2}$

It is therefore possible to find the value of $f(\alpha) / f'(\alpha)$ by dividing $f(x)$ twice by the linear factor $x - \alpha$ and finding the ratio of the remainders. This avoids the need for differentiation, and the method of the previous section can be used to manipulate the coefficients of $f(x)$, $q_1(x)$ and $q_2(x)$.

Consider the problem of finding $-f(2)/f'(2)$

$$\text{if } f(x) = 3x^3 + 2x^2 - x - 2.$$

Let array A hold the coefficients of $f(x)$, and array B the coefficients of $q_1(x)$ and the remainder r_1. A third array C can then hold the coefficients of $q_2(x)$ and remainder r_2.

A	+3	+2	-1	-2
B	+3	+8	+15	+28(=r_1)
C	+3	+24	+63(=r_2)	

Thus $-\dfrac{f(2)}{f'(2)} = -\dfrac{r_1}{r_2} = -\dfrac{28}{63}$

9.3. REAL ROOTS

Program 9.3 finds the real roots of a polynomial equation in the form

$$A[n]x^n + A[n-1]x^{n-1} + \ldots + A[1]x + A[0] = 0.$$

A first estimate for a possible root is taken as $x = 1$, and this is improved by the Newton–Raphson formula, giving

$$x = 1 + d, \quad \text{where } d = -\frac{f(1)}{f'(1)}$$

The correction factor, d, is calculated by the division process explained on the opposite page. A loop is established in lines 11–24, and the division is carried out in lines 11–14. The value of the correction factor is then assigned to d in line 16.

When the counter r records that 20 iterations have taken place, the statement

$$\underline{\text{if}} \text{ abs } (d/x) > 10{\uparrow}(-4) \text{ } \underline{\text{then}} \text{ } \underline{\text{goto}} \text{ L32;}$$

ends the program if convergence has not yet taken place. This will occur when all the real roots have been removed and only complex roots remain. It can also occur if the polynomial is not a well-behaved function.

The next statement

$$\underline{\text{if}} \text{ abs } (d) \geqslant \text{ abs (previous d) } \underline{\text{then}} \text{ } \underline{\text{goto}} \text{ L25;}$$

makes a noise-level comparison, and prints the root x if it has been found to the maximum accuracy of the computer. If the answer to both of the above questions is 'no', further iterations are performed until noise-level accuracy is achieved.

Providing the counter j shows that not all roots have been found, the next polynomial to be solved is transferred from array B to array A. A search is then made for a new root, using the previous root as its first estimate.

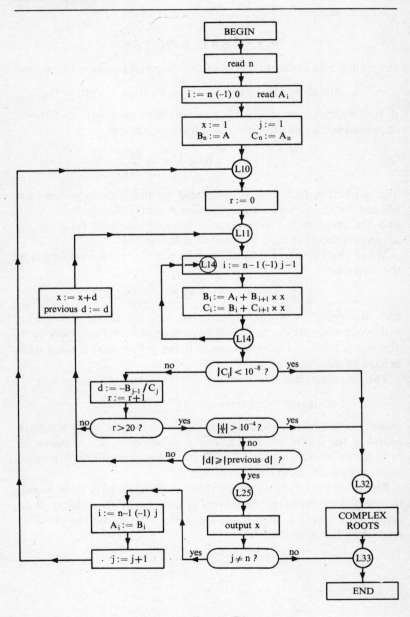

Fig. 9.3(a). Real roots.

```
begin      comment To find the real roots of a polynomial of degree n;
           integer i, j, n, r;
           real d, previous d, x;
           open (20);    open (30);
           n := read (20);
           begin     array A, B, C[0: n];
                     for i := n step −1 until 0 do A[i] := read (20);
                     x := 1;    j := 1;
                     B[n] := C[n] := A[n];
           L10:      r := 0;
           L11:      for i := n − 1 step −1 until j − 1 do
                     begin B[i] := A[i] + B[i + 1] × x;
                         | C[i] := B[i] + C[i + 1] × x;
                 L14: end;

                     if abs(C[j]) < ₁₀−8 then goto L32;
                     d := − B[j − 1]/C[j];
                     r := r + 1;
                     if r > 20 then
                     begin     if abs (d/x) > ₁₀−4 then goto L32;
                         |     if abs (d) > abs (previous d) then goto L:
                       end;
                     x := x + d;
                     previous d := d;
                     goto L11;

           L25:      output (30, x);
                     if j ≠ n then
                     begin for i := n−1 step −1 until j do A[i] := B|
                         | j := j + 1;
                         | goto L10
                       end
                     else goto L33;
           L32:      writetext (30, [COMPLEX*ROOTS]);
       L33: end;
           close (20);    close (30)
 end →
```

7; 6; 5; −125; 6; 67; 197; 60; 0; → →

Fig. 9.3(b). Real roots.

Division of $a^n x^n + \cdots + a_0$ by $b^m x^m + \cdots + b_0$ $(n \geq m)$

$$
\begin{array}{l}
 q_{n-m} x^{n-m} + q_{n-1-m} x^{n-1-m} + \cdots\cdots\cdots + q_0 \\[4pt]
b_m x^m + \cdots\cdots + b_0 \,\big)\ a_n x^n + a_{n-1} x^{n-1} + a_{n-2} x^{n-2} + \cdots\cdots + a_{n-m} x^{n-m} + \cdots\cdots + a_0 \\[2pt]
 q_{n-m} b_m x^n + q_{n-m} b_{m-1} x^{n-1} + q_{n-m} b_{m-2} x^{n-2} + \cdots\cdots + q_{n-m} b_0 x^{n-m} \\[2pt]
\hline
 a'_{n-1} x^{n-1} + a'_{n-2} x^{n-2} + \cdots\cdots + a'_{n-m} x^{n-m} + \cdots\cdots + a_0 \\[2pt]
 q_{n-1-m} b_m x^{n-1} + q_{n-1-m} b_{m-1} x^{n-2} + \cdots\cdots + q_{n-1-m} b_{n-m+1} x^{n-m} + q_{n-1-m} b_{n-m} x^{n-m-1} \\[2pt]
\hline
 a''_{n-2} x^{n-2} + \cdots\cdots + a''_{n-m} x^{n-m} + \cdots\cdots + a''_{n-m-1} x^{n-m-1} \\[6pt]
\cdot\qquad\qquad\cdot\qquad\qquad\qquad\cdot\qquad\qquad\qquad\cdot \\[6pt]
 a^*_m x^m + a^*_{m-1} x^{m-1} + \cdots\cdots + a^*_1 x + a_0 \\[2pt]
 q_0 b_m x^m + q_0 b_{m-1} x^{m-1} + \cdots\cdots + q_0 b_1 x + q_0 b_0 \\[2pt]
\hline
 r_{m-1} x^{m-1} + \cdots\cdots + r_1 x + r_0
\end{array}
$$

Fig. (iv).

9.4. POLYNOMIAL DIVISION

The procedure POLYDIV divides a polynomial A, of degree n, by a polynomial B, of degree m, to give the quotient Q and the remainder R, of degree r. The procedure follows the standard method of long division as set out in Fig. (iv).

The coefficients of the quotient are found from the sequence

$$q_{n-m} = \frac{a_n}{b_m}; \qquad q_{n-m-1} = \frac{a'_{n-1}}{b_m}; \qquad q_{n-2-m} = \frac{a''_{n-2}}{b_m};$$

which may be generalized by the statement

$$Q_{i-m} := A_i / B_m \qquad (1)$$

where the counter i represents the degree of the remainder at any stage of the division process.

The coefficients of the successive remainders are given by the relationships

$$a'_n = a_n - q_{n-m}b_m \quad (= 0) \qquad a''_{n-1} = a'_{n-1} - q_{n-1-m}b_{m-1} \quad (=$$
$$a'_{n-1} = a_{n-1} - q_{n-m}b_{m-1} \qquad a''_{n-1-1} = a'_{n-1-1} - q_{n-1-m}b_{m-2}$$
$$a'_{n-2} = a_{n-2} - q_{n-m}b_{m-2} \qquad a''_{n-1-2} = a'_{n-1-2} - q_{n-1-m}b_{m-2}$$

$$\cdots\cdots\cdots\cdots\cdots\cdots\cdots \qquad \cdots\cdots\cdots\cdots\cdots\cdots\cdots$$

$$a'_{n-j} = a_{n-j} - q_{n-m}b_{m-j} \qquad a''_{n-1-j} = a'_{n-1-j} - q_{n-1-m}b_{m-j}$$

$$\cdots\cdots\cdots\cdots\cdots\cdots\cdots \qquad \cdots\cdots\cdots\cdots\cdots\cdots\cdots$$

These relationships may be generalized by the statement

$$A_{i-j} := A_{i-j} - Q_{i-m} \times B_{m-j} \qquad (2)$$

where A_{i-j} is the coefficient of the term in x^{i-j} of the remainder. It will be noticed that for each value of i, the counter j runs from 0 to m. It is not necessary to retain the notation $a', a'', \ldots, a*$ in the assignment statement (2) because the old value of A_{i-j} is no longer required after it has been used in the right-hand side of the statement. After all the values of i and j have been considered, the array A holds the final remainder, and this is transferred to array R.

Loss of accuracy during the subtraction process is always expected, and each element A_{i-j} is set to zero if there is a loss of more than eight significant figures. The degree of the remainder is found in lines 16 and 17, so that this value may be used in any further call on the procedure. Examples on the use of this procedure are set as problems.

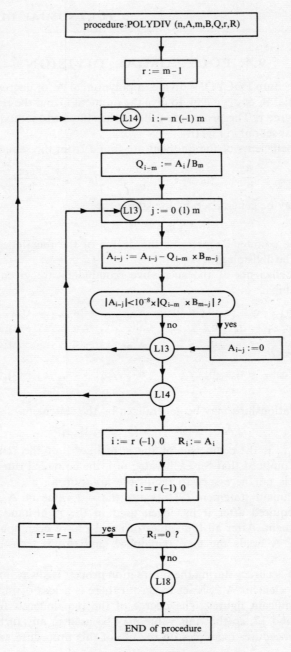

Fig. 9.4(a). Polynomial division.

162

```
procedure POLYDIV (n, A, m, B, Q, r, R);
value     n, m, A, B;
integer   n, m, r;
array     A, B, Q, R;
begin     integer i, j;
          r := m − 1;
          for i := n step −1 until m do
          begin    Q[i − m] := A[i]/B[m];
                   for j := 0 step 1 until m do
                   begin A[i−j] := A[i−j]−Q[i−m]×B[m−j];
                         if abs (A[i−j]) < ₁₀−8×abs(Q[i−m]×B[m−j])
                         then A[i − j] := 0;
          L13: end;
L14: end;
          for i := r step −1 until 0 do R[i] := A[i];
          for i := r step −1 until 0 do
          if R[i] = 0 then r := r − 1 else goto L18;
L18: end;
```

Fig. 9.4(b). Polynomial division.

PROBLEMS

1. Write a program to solve the general quadratic equation for all values of a, b and c, including a = 0, b = 0 and c = 0, using a method of error control for very small roots.

2. Write a procedure POLYMULT to find the product C of two polynomials A and B of degrees m and n.

3. Two ladders of length p and q lean against opposite sides of an alley. If h is the height at which they cross, write a program to find the width, w, of the alley. It can be shown by similarity and by Pythagoras' theorem that

$$\frac{h}{x} + \frac{h}{y} = 1 \tag{1}$$

$$x^2 + w^2 = p^2 \tag{2}$$

$$y^2 + w^2 = q^2 \tag{3}$$

163

Elimination of w and y from these equations gives a quartic in x which may be solved either by a direct or by an iterative method. Hence w can be found from equation (2).

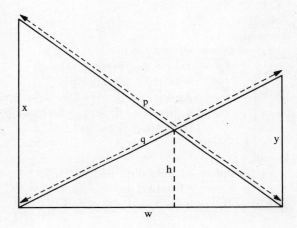

4. The highest common factor of two polynomials is the polynomial of highest degree which divides into both polynomials exactly. For example, the highest common factor of $x^2 - 2x + 1$ and $x^2 + 2x - 3$ is $x - 1$. The highest common factor of two polynomials $f(x)$ and $g(x)$ is found as follows:

(a) Divide $f(x)$ by $g(x)$, obtaining quotient $q_1(x)$ and remainder $r_1(x)$.

   Then $\qquad f(x) = g(x).q_1(x) + r_1(x).$

(b) Divide $g(x)$ by $r_1(x)$, obtaining quotient $q_2(x)$ and remainder $r_2(x)$.

   Then $\qquad g(x) = f_1(x).q_2(x) + r_2(x).$

(c) Divide $r_1(x)$ by $r_2(x)$, obtaining quotient $q_3(x)$ and remainder $r_3(x)$.

   Then $\qquad r_1(x) = r_2(x).q_3(x) + r_3(x).$

(d) Continue this process until at some stage the remainder is zero, i.e.

$$r_2(x) = r_3(x).q_4(x) + r_4(x)$$
$$\cdots\cdots\cdots\cdots\cdots\cdots\cdots\cdots\cdots$$
$$r_{n-2}(x) = r_{n-1}(x).q_n(x) + r_n(x)$$
$$r_{n-1}(x) = r_n(x).q_{n+1}(x) + 0$$

The last non-zero remainder, $r_n(x)$, is then the HCF of $f(x)$ and $g(x)$.

Write a program to find the highest common factor of two polynomials whose coefficients are given on a data tape.

164

5. Bairstow's method. If a polynomial equation has imaginary, as well as real roots, these occur in pairs and correspond to quadratic factors of the polynomial. A general polynomial equation may therefore be solved by splitting it into a set of quadratic equations which may then be solved by formula.

Bairstow's method is to take a trial quadratic expression, $x^2 + px + q$, and divide this into the given polynomial $f(x)$ to obtain a remainder. p and q are then altered to make the remainder smaller, and the process is repeated until the remainder is considered small enough. The formulae for improving the values of p and q are found as follows:

Divide $f(x)$ by $x^2 + px + q$ giving quotient

$$q_{n-2}x^{n-2} + q_{n-3}x^{n-3} + \ldots + q_1x + q_0$$

and remainder $r_1x + r_0$.

If r_1 and r_0 are zero, the quadratic expression is an exact factor, otherwise construct a new polynomial

$$b_nx^n + b_{n-1}x^{n-1} + \ldots + b_1x + b_0$$

where $\qquad b_n = q_{n-2}, \qquad b_{n-1} = q_{n-3}, \qquad \ldots, \qquad b_2 = q_0$

and $\qquad\qquad\qquad\qquad b_1 = r_1, \qquad b_0 = r_0.$

Divide this new polynomial by $x^2 + px + q$, giving quotient

$$c_{n-2}x^{n-2} + c_{n-3}x^{n-3} + \ldots + c_1x + c_0$$

and remainder $\qquad\qquad\qquad t_1x + t_0$

Improved values for p and q can then be taken as

$$p + \Delta p$$
$$q + \Delta q$$

where $\qquad\qquad \Delta p = [c_0r_1 - c_1(pr_1 + r_0)]/\lambda$

and $\qquad\qquad \Delta q = [c_0r_0 - r_1(t_1 - r_1 + pc_0)]/\lambda \qquad [\lambda \neq 0]$

where $\qquad\qquad \lambda = c_0^2 - c_1(t_1 - r_1 + 2pc_0).$

Write a program to solve a general polynomial equation by Bairstow's method, using the procedure POLYDIV and taking initial values of $p = 0$ and $q = 0$ in the trial quadratic expression, $x^2 + px + q$.

6. Write a program to solve a general polynomial equation by first extracting the real roots by the Newton–Raphson method, and then the complex roots by Bairstow's method.

10

MATRICES AND DETERMINANTS

INTRODUCTION

Notation used in this chapter

A, B, C, etc.   Matrices or arrays.

det A   The determinant of a square matrix A.

A_{ij} The element in the ith row and j^{th} column of a matrix or determinant.

A_{ij}^* The cofactor of element A_{ij} in det A.

A[i, j]   The subscripted variable used in Algol to store the value of the element A_{ij}.

A^{-1} The inverse of the matrix A.

Although a_{ij} is the standard notation for an element of a matrix, the notation A_{ij} has been used throughout this chapter to show more clearly the correspondence with A[i, j], an element of an array. The cofactor of A_{ij} is therefore denoted by A_{ij}^*.

Definitions and results

§1. A matrix having m rows and n columns is called a matrix of type m × n. If m = n, the matrix is said to be square.

§2. The product AB of two matrices is only defined if the number of rows of A is the same as the number of columns of B. The matrices are then said to be compatible, and their product C is defined by

$$C_{ij} = \sum_{\lambda} A_{i\lambda} B_{\lambda j}$$

If A is of type m × n and B is of type n × p, C is of type m × p and the summation is taken from $\lambda = 1$ to $\lambda = n$.

§3. The determinant of a square matrix

$$A = \begin{pmatrix} A_{11} & . & . & . & A_{1n} \\ . & & . & . & . \\ . & & . & . & . \\ A_{n1} & . & . & . & A_{nn} \end{pmatrix}$$

is written as $\quad \det A = \begin{vmatrix} A_{11} & . & . & . & A_{1n} \\ . & & . & . & . \\ . & & . & . & . \\ A_{n1} & . & . & . & A_{nn} \end{vmatrix}$

The determinant of a non-square matrix is not defined.

§4. Expansion of determinants

(a) The minor of an element A_{ij} of det A is found from det A by deleting the elements in the ith row and j^{th} column.

(b) The cofactor A_{ij}^* of an element A_{ij} is related to the minor of A_{ij} in the following way:

$$\text{cofactor of } A_{ij} = (-1)^{i+j} \text{ (minor of } A_{ij}).$$

(c) The expansion of a determinant of order n by the first row is defined as follows:

$$\det A = \begin{vmatrix} A_{11} & . & A_{1n} \\ . & & . \\ A_{n1} & . & A_{nn} \end{vmatrix} = A_{11}A_{11}^* + A_{12}A_{12}^* + \ldots + A_{1n}A_{1n}^*$$

(d) If the determinant is of order 2, the expansion is given by:

$$\det A = \begin{vmatrix} A_{11} & A_{12} \\ A_{21} & A_{22} \end{vmatrix} = A_{11}A_{22} - A_{12}A_{21}$$

Note that in this case, $A_{11}^* = A_{22}$ and $A_{12}^* = A_{21}$.

§5. Properties of determinants

(a) An interchange of two rows (or columns) reverses the sign.

(b) A factor common to all members of a row (or column) may be taken outside:

$$\begin{vmatrix} A_{11} & \mu A_{12} & A_{13} \\ \lambda A_{21} & \lambda\mu A_{22} & \lambda A_{23} \\ A_{31} & \mu A_{32} & A_{33} \end{vmatrix} = \lambda\mu \begin{vmatrix} A_{11} & A_{12} & A_{13} \\ A_{21} & A_{22} & A_{23} \\ A_{31} & A_{32} & A_{33} \end{vmatrix}$$

(c) The value is unaltered if any multiple of a row (or column) is added to any other row (or column):

$$\begin{vmatrix} A_{11} & A_{12} & A_{13} \\ A_{21} & A_{22} & A_{23} \\ A_{31} & A_{32} & A_{33} \end{vmatrix} = \begin{vmatrix} A_{11}+\lambda A_{21} & A_{12}+\lambda A_{22} & A_{13}+\lambda A_{23} \\ A_{21} & A_{22} & A_{23} \\ A_{31} & A_{32} & A_{33} \end{vmatrix}$$

§6. The elements A_{kk} of a square matrix or determinant are called diagonal elements, and the diagonal containing the elements A_{kk}

167

for all values of k is called the leading diagonal. If all the elements below the leading diagonal are zero, the matrix or determinant is said to be in upper triangular form.

§7. Reduction of determinants to upper triangular form

The determinant
$$\det A = \begin{vmatrix} A_{11} & A_{12} & A_{13} \\ A_{21} & A_{22} & A_{23} \\ A_{31} & A_{32} & A_{33} \end{vmatrix}$$

may be reduced to upper triangular form using the properties of §5:

(a) Multiply row 1 by $\dfrac{A_{21}}{A_{11}}$ and subtract the products from row 2.

(b) Multiply row 1 by $\dfrac{A_{31}}{A_{11}}$ and subtract the products from row 3.

The determinant is now in the form

$$\det A = \begin{vmatrix} A_{11} & A_{12} & A_{13} \\ 0 & A'_{22} & A'_{23} \\ 0 & A'_{32} & A'_{33} \end{vmatrix} \quad \text{where } A'_{22} = A_{22} - A_{12} \times \frac{A_{21}}{A_{11}}, \text{ etc.}$$

(c) Multiply row 2 by $\dfrac{A'_{32}}{A'_{22}}$ and subtract the products from row 3.

Then

$$\det A = \begin{vmatrix} A_{11} & A_{12} & A_{13} \\ 0 & A'_{22} & A'_{23} \\ 0 & 0 & A''_{33} \end{vmatrix} \quad \text{where } A''_{33} = A''_{33} - A_{23} \times \frac{A'_{32}}{A'_{22}}.$$

The final determinant is in upper triangular form, and the elements on the leading diagonal are known as pivots. The value of the determinant is easily found from the product of the pivots:

$$\det A = A_{11} \cdot A'_{22} \cdot A''_{33}$$

For example:

$$\begin{vmatrix} 6 & 3 & -9 \\ -2 & 1 & 7 \\ 4 & -1 & -2 \end{vmatrix} = \begin{vmatrix} 6 & 3 & -9 \\ 0 & 2 & 4 \\ 0 & -3 & 4 \end{vmatrix} = \begin{vmatrix} 6 & 3 & -9 \\ 0 & 2 & 4 \\ 0 & 0 & 10 \end{vmatrix}$$

$$= 6 \times 2 \times 10 = 120$$

This result may be verified by the direct expansion of the original determinant.

168

§8. The determinant of the product of two matrices is the product of their determinants: det (AB) = (det A) (det B).

§9. **Unit matrices.** The square matrices

$$I_1 = (1), \quad I_2 = \begin{pmatrix} 1 & 0 \\ 0 & 1 \end{pmatrix}, \quad I_3 = \begin{pmatrix} 1 & 0 & 0 \\ 0 & 1 & 0 \\ 0 & 0 & 1 \end{pmatrix}, \text{ etc.}$$

are called unit matrices of orders 1, 2, 3 etc. If A is a square matrix and I a unit matrix of the same order, then

$$AI = IA = A.$$

§10. **The inverse of a matrix.** If A is a square matrix and I is a unit matrix of the same order, the square matrix B such that

$$AB = BA = I$$

is called the inverse of A, and is written A^{-1}. The inverse of A does not exist if det A = 0 because

$$AB = I$$
$$\Rightarrow (\det A)(\det B) = \det I \quad \text{(from §8)}$$

Since det I = 1, the last equation is not true if det A = 0. If this is the case, the matrix A is called singular. For example, matrices

$$\begin{pmatrix} 12 & 3 \\ -8 & -2 \end{pmatrix} \quad \text{and} \quad \begin{pmatrix} -1 & 2 \\ -2 & 4 \end{pmatrix}$$

are singular because their determinants are zero, whereas matrices

$$\begin{pmatrix} 1 & 2 & -1 \\ 2 & 3 & 0 \\ 1 & 1 & 4 \end{pmatrix} \quad \text{and} \quad \begin{pmatrix} 1 & 4 \\ 3 & 2 \end{pmatrix}$$

are non-singular.

§11. The inverse of a non-singular matrix may be calculated directly using the following result:

If $A = \begin{pmatrix} A_{11} & A_{12} & A_{13} \\ A_{21} & A_{22} & A_{23} \\ A_{31} & A_{32} & A_{33} \end{pmatrix}$ then $A^{-1} = \dfrac{1}{\det A} \begin{pmatrix} A_{11}^* & A_{21}^* & A_{31}^* \\ A_{12}^* & A_{22}^* & A_{32}^* \\ A_{13}^* & A_{23}^* & A_{33}^* \end{pmatrix}$

Note that the cofactors in A^{-1} are found from the elements of A after the rows and columns have been interchanged.

For example, if $A = \begin{pmatrix} 1 & 4 \\ 3 & 2 \end{pmatrix}$ then det $A = 1 \times 2 - 3 \times 4 = -10$

and therefore $\quad A^{-1} = \dfrac{1}{-10} \begin{pmatrix} 2 & -4 \\ -3 & 1 \end{pmatrix} = \begin{pmatrix} -\dfrac{1}{5} & \dfrac{2}{5} \\ \dfrac{3}{10} & -\dfrac{1}{10} \end{pmatrix}$

§12. Matrix equations and inversion by successive elimination

The equations

$$A_{11} x_1 + A_{12} x_2 + A_{13} x_3 = y_1$$
$$A_{21} x_1 + A_{22} x_2 + A_{23} x_3 = y_2$$
$$A_{31} x_1 + A_{32} x_2 + A_{33} x_3 = y_3$$

may be written in the form

$$\begin{pmatrix} A_{11} & A_{12} & A_{13} \\ A_{21} & A_{22} & A_{23} \\ A_{31} & A_{32} & A_{33} \end{pmatrix} \begin{pmatrix} x_1 \\ x_2 \\ x_3 \end{pmatrix} = \begin{pmatrix} 1 & 0 & 0 \\ 0 & 1 & 0 \\ 0 & 0 & 1 \end{pmatrix} \begin{pmatrix} y_1 \\ y_2 \\ y_3 \end{pmatrix} \qquad (1)$$

or

$$A X = I Y \qquad (2)$$

where A, X, I and Y represent the matrices in equation (1).

Premultiplying both sides of equation (2) by the inverse of A gives

$$A^{-1} A X = A^{-1} I Y \qquad (3)$$

Using $A^{-1} A = I$ and $A^{-1} I = A^{-1}$, equation (3) reduces to

$$I X = A^{-1} Y \qquad (4)$$

Equation (4) finally reduces to

$$X = A^{-1} Y$$

or $\qquad \begin{pmatrix} x_1 \\ x_2 \\ x_3 \end{pmatrix} = A^{-1} \begin{pmatrix} y_1 \\ y_2 \\ y_3 \end{pmatrix}$

from which the values of x_1, x_2 and x_3 may be deduced.

In practice, equation (2) may be reduced to equation (4) by performing the same row operations to matrix A and matrix I simultaneously. Then, as A is reduced to the identity matrix, I will be reduced to the required inverse A^{-1}. This process may be illustrated by means of an elementary example showing the effects of the row operations on the original equations.

Row operation	Equation	Matrix form
row r_1 row r_2	$3x + 2y = 4$ $2x - y = 5$	$\begin{pmatrix} 3 & 2 \\ 2 & -1 \end{pmatrix}\begin{pmatrix} x \\ y \end{pmatrix} = \begin{pmatrix} 1 & 0 \\ 0 & 1 \end{pmatrix}\begin{pmatrix} 4 \\ 5 \end{pmatrix}$
I $\begin{cases} r_1 := r_1/3 \\ r_2 := r_2 \end{cases}$	$x + \frac{2}{3}y = \frac{4}{3}$ $2x - y = 5$	$\begin{pmatrix} 1 & \frac{2}{3} \\ 2 & -1 \end{pmatrix}\begin{pmatrix} x \\ y \end{pmatrix} = \begin{pmatrix} \frac{1}{3} & 0 \\ 0 & 1 \end{pmatrix}\begin{pmatrix} 4 \\ 5 \end{pmatrix}$
II $\begin{cases} r_1 := r_1 \\ r_2 := r_2 - 2r_1 \end{cases}$	$x + \frac{2}{3}y = \frac{4}{3}$ $0x - \frac{7}{3}y = \frac{7}{3}$	$\begin{pmatrix} 1 & \frac{2}{3} \\ 0 & -\frac{7}{3} \end{pmatrix}\begin{pmatrix} x \\ y \end{pmatrix} = \begin{pmatrix} \frac{1}{3} & 0 \\ -\frac{2}{3} & 1 \end{pmatrix}\begin{pmatrix} 4 \\ 5 \end{pmatrix}$
III $\begin{cases} r_1 := r_1 \\ r_2 := r_2/-\frac{7}{3} \end{cases}$	$x + \frac{2}{3}y = \frac{4}{3}$ $0x + y = -1$	$\begin{pmatrix} 1 & \frac{2}{3} \\ 0 & 1 \end{pmatrix}\begin{pmatrix} x \\ y \end{pmatrix} = \begin{pmatrix} \frac{1}{3} & 0 \\ \frac{2}{7} & \frac{3}{7} \end{pmatrix}\begin{pmatrix} 4 \\ 5 \end{pmatrix}$
IV $\begin{cases} r_1 := r_1 - \frac{2}{3}r_2 \\ r_2 := r_2 \end{cases}$	$x + 0y = 2$ $0x + y = -1$	$\begin{pmatrix} 1 & 0 \\ 0 & 1 \end{pmatrix}\begin{pmatrix} x \\ y \end{pmatrix} = \begin{pmatrix} \frac{1}{7} & \frac{2}{7} \\ \frac{2}{7} & \frac{3}{7} \end{pmatrix}\begin{pmatrix} 4 \\ 5 \end{pmatrix}$

Finally,

$$x = 2 \atop y = -1 \quad \text{or} \quad \begin{pmatrix} x \\ y \end{pmatrix} = \begin{pmatrix} 2 \\ -1 \end{pmatrix}$$

Steps I and III reduce the elements on the leading diagonal to unity. If these elements are initially zero, a suitable multiple of some other row may be added to the row containing a zero diagonal element. Step II reduces to zero the term(s) below the leading diagonal. Step IV reduces to zero the term(s) above the leading diagonal.

§13. Cramer's rule for the solution of linear equations

The set of linear equations

$$A_{11}x_1 + A_{12}x_2 + A_{13}x_3 = B_1$$
$$A_{21}x_1 + A_{22}x_2 + A_{23}x_3 = B_2$$
$$A_{31}x_1 + A_{32}x_2 + A_{33}x_3 = B_3$$

has a unique solution in the form

$$x_1 = \frac{1}{d} \begin{vmatrix} B_1 & A_{12} & A_{13} \\ B_2 & A_{22} & A_{23} \\ B_3 & A_{32} & A_{33} \end{vmatrix} \quad x_2 = \frac{1}{d} \begin{vmatrix} A_{11} & B_1 & A_{13} \\ A_{21} & B_2 & A_{23} \\ A_{31} & B_3 & A_{33} \end{vmatrix} \quad x_3 = \frac{1}{d} \begin{vmatrix} A_{11} & A_{12} & B_1 \\ A_{21} & A_{22} & B_2 \\ A_{31} & A_{32} & B_3 \end{vmatrix}$$

where
$$d = \begin{vmatrix} A_{11} & A_{12} & A_{13} \\ A_{21} & A_{22} & A_{23} \\ A_{31} & A_{32} & A_{33} \end{vmatrix}$$

The solution may be written in the form

$$x_j = (d_j/d) \quad (j = 1, 2, 3)$$

where the determinant d_j is obtained from d by replacing the jth column of d by the column of B's on the right-hand side of the equation. If d = 0, the equations have more than one solution or are inconsistent. Cramer's rule may be extended to find the unique solution of a set of n linear equations in n unknowns.

§14. Accuracy and ill-conditioning

Considerable round-off errors can be introduced when expanding determinants or making row operations on matrices. Cases where rounding errors lead to a loss of significant figures are called ill-conditioned.

Two illustrations are given below.

(a) The example in §12 is well-conditioned and, if all the intermediate calculations are rounded to two significant figures, the resulting matrix equation would be

$$\begin{pmatrix} 1 & 0 \\ 0 & 1 \end{pmatrix} \begin{pmatrix} x \\ y \end{pmatrix} = \begin{pmatrix} 0.14 & 0.29 \\ 0.29 & -0.43 \end{pmatrix} \begin{pmatrix} 4 \\ 5 \end{pmatrix} = \begin{pmatrix} 2.0 \\ -0.99 \end{pmatrix}$$

The results are very close to the exact solution x = 2, y = −1.

If, however, the same method is applied to the equations

$$9x - 10y = 2$$
$$8x - 9y = 1$$

the solutions obtained are

$$x = 4.3, \quad y = 3.8$$

compared with the correct values of

$$x = 8.0, \quad y = 7.0$$

The steps in the solution may be illustrated as follows

$$\begin{pmatrix} 9 & -10 \\ 8 & -9 \end{pmatrix}\begin{pmatrix} x \\ y \end{pmatrix} = \begin{pmatrix} 1 & 0 \\ 0 & 1 \end{pmatrix}\begin{pmatrix} 2 \\ 1 \end{pmatrix}$$

I $\quad \begin{pmatrix} 1 & -1.1 \\ 0 & -0.2 \end{pmatrix}\begin{pmatrix} x \\ y \end{pmatrix} = \begin{pmatrix} 0.11 & 0 \\ -0.88 & 1 \end{pmatrix}\begin{pmatrix} 2 \\ 1 \end{pmatrix}$ Error caused by putting $\frac{10}{9} = 1.1$

II $\quad \begin{pmatrix} 1 & -1.1 \\ 0 & -0.2 \end{pmatrix}\begin{pmatrix} x \\ y \end{pmatrix} = \begin{pmatrix} 0.11 & 0 \\ -0.88 & 1 \end{pmatrix}\begin{pmatrix} 2 \\ 1 \end{pmatrix}$ No error

III $\quad \begin{pmatrix} 1 & -1.1 \\ 0 & 1 \end{pmatrix}\begin{pmatrix} x \\ y \end{pmatrix} = \begin{pmatrix} 0.11 & 0 \\ 4.4 & -5 \end{pmatrix}\begin{pmatrix} 2 \\ 1 \end{pmatrix}$ No error

IV $\quad \begin{pmatrix} 1 & 0 \\ 0 & 1 \end{pmatrix}\begin{pmatrix} x \\ y \end{pmatrix} = \begin{pmatrix} 4.9 & -5.5 \\ 4.4 & -5 \end{pmatrix}\begin{pmatrix} 2 \\ 1 \end{pmatrix}$ Further errors caused by putting $(4.4) \times (1.1) = 4.8$ and $0.11 + 4.8 = 4.9$

Finally,
$$\begin{pmatrix} x \\ y \end{pmatrix} = \begin{pmatrix} 4.3 \\ 3.8 \end{pmatrix}$$

(b) Consider the use of Cramer's Rule to solve the linear equations

$$19x_1 - 13x_2 = 22$$
$$39x_1 - 27x_2 = 43$$

using a machine with a maximum accuracy of two significant decimal digits. The solution, with each calculation rounded to two significant figures, is given by

$$x_1 = \frac{\begin{vmatrix} 22 & -13 \\ 43 & -27 \end{vmatrix}}{\begin{vmatrix} 19 & -13 \\ 39 & -27 \end{vmatrix}} = \frac{-590 + 560}{-510 + 510} = \frac{-30}{0}$$

and

$$x_2 = \frac{\begin{vmatrix} 19 & 22 \\ 39 & 43 \end{vmatrix}}{\begin{vmatrix} 19 & -13 \\ 39 & -27 \end{vmatrix}} = \frac{820 - 860}{-510 + 510} = \frac{-40}{0}$$

173

In each case, the machine overflows, and the solution is unobtainable. An exact working of the problem shows the solution to be

$$x_1 = 5\tfrac{5}{6}$$
and
$$x_2 = 6\tfrac{5}{6}$$

10.1. MATRIX MULTIPLICATION

The flow diagram in Fig. 10.1 describes a procedure for matrix multiplication. The matrices A and B are of types m × n and p × q. If these matrices are incompatible (i.e. n ≠ p) a warning notice is printed: otherwise B is pre-multiplied by A and the result is placed in matrix C of type m × q.

It will be noticed that C_{ij} occurs on the right-hand side of an assignment statement, and that it is set to zero before use. A procedure may be used at any time, and so it cannot be assumed that the storage space reserved for the array C has not been used before. The cells of array C must therefore be assigned a zero value before their use in the procedure.

10.2. DETERMINANTS (I)

The real procedure DETERM I is used to find the expansion by the first row of a determinant, using the formula from §4 of the introduction:

$$\det A = A_{11} A_{11}^* + A_{12} A_{12}^* + \ldots + A_{1n} A_{1n}^*$$

The array A holds the original determinant of order n, and is evaluated directly if n = 2. If n is greater than 2, the cofactors are isolated and placed in the array B of order n − 1. The value of each minor of the elements A_{1k} of the first row is found by a recursive call on the procedure, (see program 7.2), and the corresponding cofactor is found by multiplying by $(-1)^{k+1}$. The identifier d holds the sum of the products of the top row elements and their cofactors, and its final value is assigned to DETERM I.

For each value of k, the array B is filled with the elements of the array A as follows:

(a) Each element A_{ij} to the left of column k is placed in B one row higher as $B_{i-1, j}$.

174

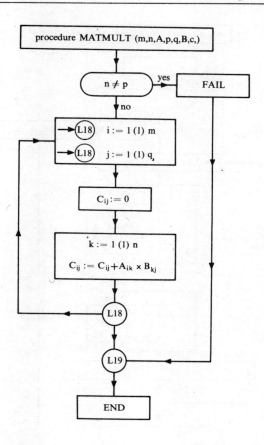

Fig. 10.1. Matrix multiplication.

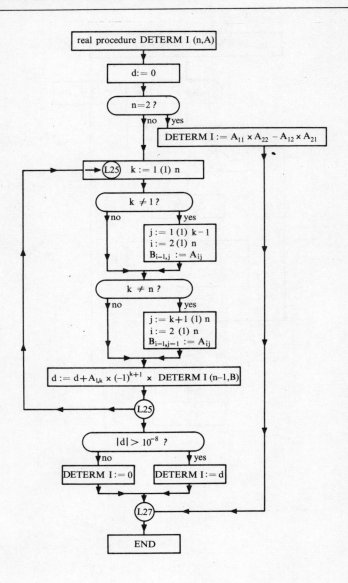

Fig. 10.2(a). Determinants (I).

```
real procedure DETERM I (n, A);
value n, A;
integer n;
array A;
comment A procedure to evaluate the determinant A by direct
expansion;
begin    array B[1: n−1, 1: n−1];
         integer i, j, k;
         real d;
         d := 0;
         if n = 2 then
         begin DETERM I := A[1, 1]×A[2, 2]−A[1, 2]×A[2, 1];
               goto L27
            end;

         for k := 1 step 1 until n do
         begin if k ≠ 1 then
               for j := 1 step 1 until k − 1 do
               for i := 2 step 1 until n do
               B[i − 1, j] := A[i, j];
               if k ≠ n then
               for j := k + 1 step 1 until n do
               for i := 2 step 1 until n do
               B[i − 1, j − 1] := A[i, j];
               d := d+A[1, k]×(−1)↑(k+1)×DETERM I (n−1,
L25: end;
         DETERM I := if abs (d) > 10−8 then d else 0;
L27: end;
```

Fig. 10.2(b). Determinants (I).

(b) Each element A_{ij} to the right of column k is placed in B one row higher and one column to the left as $B_{i-1,\,j-1}$.

(c) Row 1 and column k are not transferred to B.

A_{11}	A_{12}	A_{13}	A_{14}	A_{15}
A_{21}	A_{22}	A_{23}	A_{24}	A_{25}
A_{31}	·	·	·	·
A_{41}	·	·	·	·
A_{51}	·	·	·	·

Array A

A_{21}	A_{22}	A_{24}	A_{25}
A_{31}	A_{32}	A_{34}	A_{35}
A_{41}	·	·	·
A_{51}	·	·	·

Array B (holding the minor of A_{13})

If $k = 1$, the first operation is not necessary, nor is the second if $k = n$. In Algol, there is no objection to the instruction

$$\underline{\text{for}}\ j := 1\ \underline{\text{step}}\ 1\ \underline{\text{until}}\ 0\ \underline{\text{do}}$$

because no action will be taken. In most other languages, however, the statement is inadmissible, and the clauses

$$\underline{\text{if}}\ k \neq 1 \quad \text{and} \quad \underline{\text{if}}\ k \neq n$$

are included as a precaution.

At the end of the program, the clause

$$\underline{\text{if}}\ \text{abs}\ (d) >\ _{10}-8\ \underline{\text{then}}\ d\ \underline{\text{else}}\ 0$$

allows for errors accumulating during the expansion. If the determinant is ill-conditioned, the result should be treated with caution.

10.3. COFACTORS

The flow diagram describes the real procedure COFACT (n, A, p, q), and it is assumed that procedure DETERM I is available for use in any program in which a call on COFACT is made. The procedure uses the parameters n, A, p and q, all of which are constant in value for any particular call on the procedure. The subscripts i and j are variables local to the procedure body and must not be confused with the parameters p and q.

The first part of the flow diagram fills the top two quadrants of the array B, while lines 8–14 fill the bottom two quadrants.

178

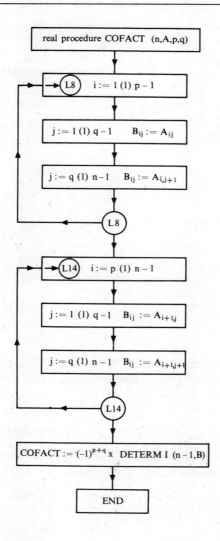

Fig. 10.3. Cofactors.

$$
\begin{array}{ccc|c|ccc}
A_{11} & \cdot & A_{1,q-1} & A_{1q} & A_{1,q+1} & \cdot & A_{1n} \\
\cdot & & \cdot & \cdot & \cdot & & \cdot \cdot \\
A_{p-1,1} & \cdot & A_{p-1,q-1} & \cdot & A_{p-1,q+1} & \cdot & A_{p-1,n} \\
\hline
A_{p,1} & \cdot & A_{p,q-1} & A_{pq} & A_{p,q+1} & \cdot & \cdot \\
\hline
A_{p+1,1} & \cdot & A_{p+1,q-1} & \cdot & A_{p+1,q+1} & \cdot & A_{p+1,n} \\
\cdot & & \cdot & \cdot & & & \\
A_{n,1} & \cdot & A_{n,q-1} & \cdot & A_{n,q+1} & \cdot & A_{nn}
\end{array}
$$

Array A

$$
\begin{array}{ccc|ccc}
A_{11} & \cdot & A_{1,q-1} & A_{1,q+1} & \cdot & A_{1n} \\
\cdot & & \cdot & \cdot & & \cdot \cdot \\
A_{p-1,1} & \cdot & A_{p-1,q-1} & A_{p-1,q+1} & \cdot & A_{p-1,n} \\
\hline
A_{p+1,1} & \cdot & A_{p+1,q-1} & A_{p+1,q+1} & \cdot & A_{p+1,n} \\
\cdot & & \cdot & \cdot & & \cdot \cdot \\
A_{n,1} & \cdot & A_{n,q-1} & A_{n,q+1} & \cdot & A_{nn}
\end{array}
$$

Array B

(holding the minor of A_{pq})

The cofactor is then evaluated by inserting the appropriate sign before the determinant of B. Use is made of this procedure in the following program.

10.4. MATRIX INVERSION

This procedure uses the result in §11 of the introduction, and finds the inverse of a matrix A from the formula

$$
A^{-1} = \frac{1}{\det A}
\begin{pmatrix}
A_{11}^{*} & \cdot & A_{n1}^{*} \\
\cdot & \cdot & \cdot \\
A_{1n}^{*} & \cdot & A_{nn}^{*}
\end{pmatrix}
$$

It employs the procedures DETERM I and COFACT, both of which must be declared at the head of any program calling upon MATINV. The statement

$$B[j, i] := \text{COFACT}(n, A, i, j)/d;$$

carries out the necessary transposition of the elements of A, by

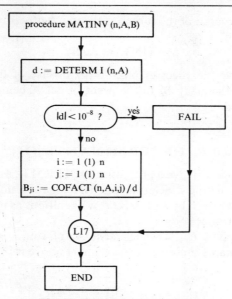

Fig. 10.4(a). Matrix inversion.

```
procedure  MATINV (n, A, B);
value      n, A;
integer    n;
array      A, B;
comment    A procedure to find the inverse, B, of the matrix A,
           using cofactors;
begin      integer i, j;
           real d;
           d := DETERM I (n, A);

           if abs(d) < ₁₀−8 then
           begin    writetext (30, [FAIL]);
                    goto L17
              end;

           for i := 1 step 1 until n do
           for j := 1 step 1 until n do
           B[j, i] := COFACT (n, A, i, j)/d;
L17: end;
```

Fig. 10.4(b). Matrix inversion.

181

means of the reversal of the subscripts i and j. A procedure call would take the form of

$$\text{MATINV (n, A, B);}$$

and would have the effect of placing the inverse of the matrix A, of order n, in the array B. In the event of det A being zero, a fail notice is printed.

10.5. DETERMINANTS (II)

Figs. 10.5 (a) and (b) give a procedure, DETERM II, as an improved method of evaluating a determinant. Here, a reduction is made to upper triangular form by successive row operations as in the example of §7 of the introduction.

For every pivot A_{kk}, each element A_{ij} of row i is replaced by itself less a suitable multiple of the element above it in row k.

$$
\begin{vmatrix}
\cdot & \cdot & \cdot & \cdot & \cdot & \cdot \\
\cdot & A_{kk} & \cdot & \cdot & A_{kj} & \cdot \\
\cdot & \cdot & \cdot & \cdot & \cdot & \cdot \\
\cdot & A_{ik} & \cdot & \cdot & A_{ij} & \cdot \\
\cdot & \cdot & \cdot & \cdot & \cdot & \cdot
\end{vmatrix}
=
\begin{vmatrix}
\cdot & \cdot & \cdot & \cdot & \cdot & \cdot \\
\cdot & A_{kk} & \cdot & \cdot & A_{kj} & \cdot \\
\cdot & \cdot & \cdot & \cdot & \cdot \\
\cdot & (0) & \cdot & \cdot & A_{ij} - A_{kj} \times \dfrac{A_{ik}}{A_{kk}} & \cdot \\
\cdot & \cdot & \cdot & \cdot & \cdot
\end{vmatrix}
$$

$$(k = 2)$$

This multiple is always $\dfrac{A_{ik}}{A_{kk}}$ so that A_{ik} is itself reduced to zero:

$$A_{ij} := A_{ij} - A_{kj} \times \frac{A_{ik}}{A_{kk}} \quad (= 0 \text{ when } j = k)$$

In practice, the elements of column k are not actually replaced by zeros because they play no further part in the calculation. Instead only values of i and j greater than k are considered.

To improve the accuracy of the solution, the largest element is made the pivot, A_{kk}, of each successive determinant, and lines 12–17 find the row p and column q of this element. If the largest element is not in the row or column of the pivot, lines 18–27 effect the necessary interchange of rows and columns using the identifier, a, as a temporary store location, and changing the sign of a row or column each time a transfer is made.

The identifier, d, initially unity, is multiplied by each pivot in turn and, if it becomes less than 10^{-8} in absolute value, the determinant is taken to be zero; otherwise reduction to upper triangular form is continued until $k = n - 1$ and the determinant is in the form:

$$\begin{vmatrix} A_{11} & A_{12} & A_{13} & . & A_{1,\,n-1} & A_{1,\,n} \\ 0 & A_{22} & A_{23} & . & A_{2,\,n-1} & A_{2,\,n} \\ 0 & 0 & A_{33} & . & A_{3,\,n-1} & A_{3,\,n} \\ . & . & . & . & . & . \\ 0 & 0 & 0 & . & A_{n-1,\,n-1} & A_{n-1,\,n} \\ 0 & 0 & 0 & . & 0 & A_{n,\,n} \end{vmatrix}$$

d is now the product of A_{11}, A_{22}, ..., $A_{n-1,\,n-1}$ and needs only to be multiplied by A_{nn} before being assigned as the final value of the determinant.

Line 35 tests each of the subtractions in line 34, to see if more than eight significant figures are lost and, if so, A_{ij} is assigned the value of zero. Of the two methods given for evaluating determinants, an advantage of program 10.2 is that fewer subtractions of nearly equal numbers are involved. It is, however, the less sophisticated of the two and a brute force expansion of this nature is comparatively slow because a total of n! determinants must be evaluated if the original order is n.

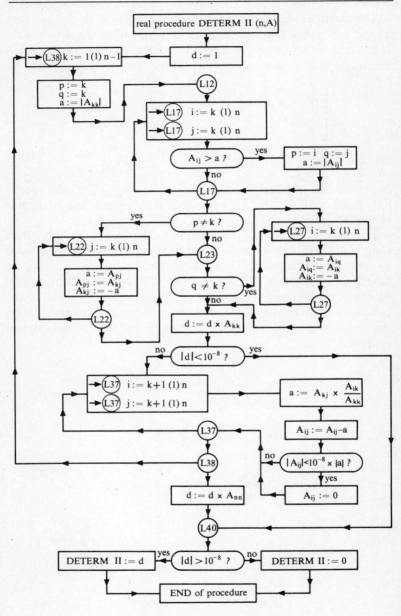

Fig. 10.5(a). Determinants (II).

```
real procedure DETERM II (n, A);
value    n, A;
integer  n;
array    A;
comment A procedure to evaluate a determinant by the leading diagonal
         method, using largest pivots;
begin    integer i, j, k, p, q;
         real a, d;
         d := 1;
         for k := 1 step 1 until n − 1 do
         begin    p := q := k;   a := abs (A[k, k]);
         L12:     for i := k step 1 until n do
                  for j := k step 1 until n do if abs (A[i, j]) > a then
                  begin    p := i;
                           q := j;
                           a := abs (A[i, j]);
         L17: end         of search for largest element;

                  if p ≠ k then for j := k step 1 until n do
                  begin    a := A[p, j];
                           A[p, j] := A[k, j];
                           A[k, j] := −a;
         L22: end         of interchange of rows p and k;

                  if q ≠ k then for i := k step 1 until n do
                  begin    a := A[i, q];
                           A[i, q] := A[i, k];
                           A[i, k] := −a;
         L27: end         of interchange of columns q and k.
                           Largest element is now the pivot;

                  d := d × A[k, k];
                  if abs(d) < ₁₀−8 then goto L40;
                  for i := k + 1 step 1 until n do
                  for j := k + 1 step 1 until n do
                  begin    a := A[k, j] × A[i, k]/A[k ,k];
                           A[i, j] := A[i, j] − a;
                           if abs (A[i, j]) < ₁₀−8 × abs(a)
                           then A[i, j] := 0;
         L37: end         of reduction to upper triangular form;
         L38: end;

         d := d × A[n, n];
L40:     DETERM II := if abs(d) > ₁₀−8 then d else 0;
end;
```

Fig. 10.5(b). Determinants (II).

185

10.6. LINEAR EQUATIONS

The final program solves a set of n linear equations in n unknowns using the method of successive elimination shown in the following example.

The equations

$$x + \ y + 2z = \ \ 3 \qquad (1)$$
$$2x - \ y + 3z = \ \ 1 \qquad (2)$$
$$x + 4y - 2z = -2 \qquad (3)$$

may be solved by eliminating x from equation (2) and x and y from equation (3):

$$x + y + 2z = \ \ 3 \qquad (4)$$
$$-3y - \ z = -5 \qquad (5)$$
$$-5z = -10 \qquad (6)$$

The value of z may now be found directly from equation (6), and the values of y and x by back substitution in equations (5) and (4). The solution is then found to be $(-2, 1, 2)$.

Fig. 10.6(b) is given in the form of a procedure, and the coefficients are stored in an array A with a bound-pair-list of $[1 : n, 1 : n + 1]$. Thus the above equations are stored as

$$A = \begin{pmatrix} 1 & 1 & 2 & 3 \\ 2 & -1 & 3 & 1 \\ 1 & 4 & -2 & -2 \end{pmatrix}$$

The successive elimination is carried out by row and column operations on the array A using maximum pivots as described in the previous program. Lines 12–23 are similar in both programs and are omitted from the flow diagram in Fig. 10.6(a). The only differences are that no sign change is made in line 21 and the <u>for</u> statement in line 18 runs from 1 to n + 1.

The interchange of rows (lines 18–22) does not affect the solution of the equations. It is therefore unnecessary to change the sign of the coefficients in the row containing the pivot. However, the interchange of columns in lines 23–28 alters the order in which the unknowns are evaluated. For example the element of largest modulus in array A is found to be A_{32} ($p = 3$ and $q = 2$). This element may be transferred to the top row by a direct interchange of rows 1 and 3.

$$A = \begin{pmatrix} 1 & 4 & -2 & -2 \\ 2 & -1 & 3 & 1 \\ 1 & 1 & 2 & 3 \end{pmatrix}$$

If columns 1 and 2 are now interchanged, and the equations solved as before, the solutions would be found as

$$(1, -2, 2)$$

instead of the required

$$(-2, 1, 2).$$

In order that the solutions may be printed in the correct order, a supplementary array, B, is employed to record the columns in which the coefficients of the three unknowns are stored. The two figures below show the values stored in array B before and after the interchange of columns 1 and 2.

	Column numbers			
Array B	1	2	3	
Array A	1	4	-2	-2
	2	-1	3	1
	1	1	2	3

	Column numbers			
	2	1	3	
	4	1	-2	-2
	-1	2	3	1
	1	1	2	3

Column numbers 1, 2 and 3 correspond to the unknowns x, y and z respectively.

Lines 33–40 carry out the reduction to upper triangular form of the matrix of coefficients of the left-hand sides of the equations. As in earlier programs, loss of accuracy is expected, and elements are assumed to be zero if more than eight significant figures are lost during the subtraction process.

If any pivot is found to be zero, the determinant of the coefficients of the unknowns is singular. In this case, no unique solution exists, and the procedure is ended either by the escape clause in line 33 or, if the last element A_{nn} is zero, by the clause in line 42.

187

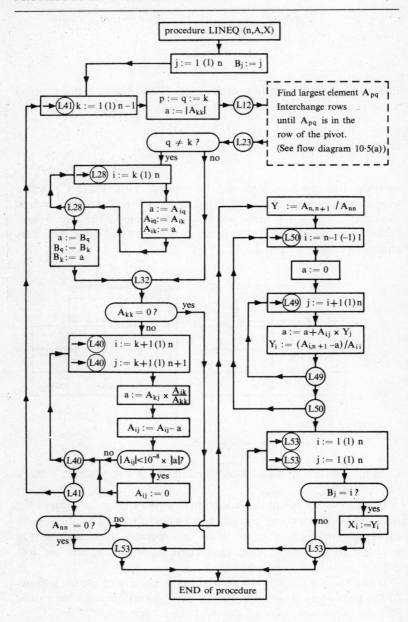

Fig. 10.6(a). Linear equations.

```
procedure LINEQ (n, A, X);
value n, A̅;
integer n;
array̅ A, X;
begin    integer   i, j, k, p, q;
         real      a;
         array     Y, B[1 : n];
         for j := 1 step 1 until n do B[j] := j;
         for k := 1 step 1 until n − 1 do
         begin    p := q := k;
                  a := abs(A[k, k]);
      L12:        for i := k step 1 until n do
                  for j := k step 1 until n do if abs(A[i, j]) > a then
                  begin    p := i;
                           q := j;
                           a := abs(A[i, j]);
              L17: end      of search for largest element;
                  if p ≠ k then for j := k step 1 until n + 1 do
                  begin    a := A[p, j];
                           A[p, j] := A[k, j];
                           A[k, j] := a;
              L22: end      of interchange of rows p and k;
                  if q ≠ k then
                  begin    for i := k step 1 until n do
                           begin    a := A[i, q];
                                    A[i, q] := A[i, k];
                                    A[i, k] := a;
                       L28: end      Largest element is now the pivot;
                           a := B[q];
                           B[q] := B[k];
                           B[k] := a;
              L32: end      of interchange of columns q and k;
                  if   A[k, k] = 0 then goto L53;
                  for i := k + 1 step 1 until n do
                  for j := k + 1 step 1 until n + 1 do
                  begin    a := A[k, j] × A[i, k]/A[k, k];
                           A[i, j] := A[i, j] − a;
                           if abs (A[i, j]) < 10−8 × abs(a)
                           then A[i, j] := 0;
              L40: end;
      L41: end;

         if A[n, n] = 0 then goto L53;
         Y[n] := A[n, n + 1]/A[n, n];
         for i := n − 1 step −1 until 1 do
         begin    a := 0;
                  for j := i + 1 step 1 until n do
                  begin    a := a + A[i, j] × Y[j];
                           Y[i] := (A[i, n + 1] −a)/A[i ,i];
              L49: end;
      L50: end    of solution of equations;
         for i := 1 step 1 until n do
         for j := 1 step 1 until n do if B[j] = i then X[i] := Y[j];
L53: end     of sorting solutions into correct order;
```

Fig. 10.6(b). Linear equations.

189

The array Y[1: n] is used to store the values of the unknowns found in lines 43–50. At the end of the procedure, the statement

for i := 1 step 1 until n do
for j := 1 step 1 until n do
if B[j] = i then X [i] := Y [j];

uses the information stored in array B to unscramble the solutions in array Y and to transfer them in the correct order into array X.

PROBLEMS

1. Devise a procedure MATWRITE (f, d, m, n, A) which will print out an array A with dimensions m × n in a given format, f, on the output device, d.

2. Write a program to show that the multiplication of matrices A, B and C of types m × n, n × p, p × q respectively is associative, i.e. that the products

$$(A\,B)\,C \quad \text{and} \quad A\,(B\,C)$$

give rise to the same matrix. Use procedure MATWRITE from problem 1 to print these products.

3. This problem illustrates that the commutative law is obeyed by a matrix and its inverse, and that

$$A\,A^{-1} = A^{-1}\,A = I.$$

Use the procedures MATINV, DETERM II, COFACT and MATMULT, and MATWRITE from problem 1, to output the matrices

$$A, \quad A^{-1}, \quad A\,A^{-1} \quad \text{and} \quad A^{-1}\,A.$$

4. Write a program to solve a set of n linear equations in n unknowns using Cramer's rule.

5. Use a procedure for matrix inversion to solve a set of n linear equations in n unknowns. Use the property that

$$A\,X = B \quad \Rightarrow \quad X = A^{-1}\,B.$$

6. The rank of a matrix. From the elements common to r given columns and r given rows of a square matrix of order n, it is possible to construct a determinant of order r. By altering the choice of rows and columns, a set of determinants of order r may be obtained, some of which may have value zero. The rank of a matrix is defined as the value of r for which at least one such determinant is non-zero and yet all determinants of order

r + 1 are zero. Rank is preserved by elementary row and column operations. The rank is most easily recognized if the matrix is transformed into
upper triangular form using the method of maximum pivots. In this case,
the rank is the number of non-zero terms counting down the leading
diagonal. Thus the ranks of

$$\begin{pmatrix} 7 & 1 & 2 \\ 0 & 4 & 1 \\ 0 & 0 & 2 \end{pmatrix} \text{ and } \begin{pmatrix} 9 & 1 & 4 & 2 \\ 0 & 3 & 1 & 2 \\ 0 & 0 & 2 & 1 \\ 0 & 0 & 0 & 0 \end{pmatrix} \text{ and } \begin{pmatrix} 4 & 2 & 0 \\ 0 & 0 & 1 \\ 0 & 0 & 0 \end{pmatrix} \text{ and } \begin{pmatrix} 1 & 0 \\ 0 & 0 \end{pmatrix}$$

are 3, 3, 1, 1.

Write a procedure to find the rank of a square matrix of order n.

7. Write a procedure to find the inverse of a matrix using the method of
§12 of the introduction. Use elementary row operations to reduce the
matrix A first to upper triangular form and then to an identity matrix.
The same operations should be used to convert an array B, initially
holding an identity matrix, into the inverse of A. In order to avoid division
by zero, a row containing a zero pivot should be replaced by itself plus
another suitable row. If no such row exists, and the pivot cannot thereby
be made non-zero, the matrix is singular and an escape clause is required.
Note that, as in program 10.5, operations need not be carried out on the
column of the pivot itself. Similarly, once A has been reduced to upper
triangular form with ones along the leading diagonal, the operations necessary to turn it into an identity matrix need only be applied to the array B.

8. Write a program to improve the accuracy of an inverse obtained by
elementary row operations. Let the array C contain I − AB, where A
is a matrix, B its approximate inverse and I the identity matrix. If C is a
zero matrix, then B is already an accurate inverse. If C is non-zero and
each element is less than 1 in absolute value, the accuracy of the inverse
may be improved by replacing B by B × (2I − AB). The accuracy of the
new approximation to the inverse of A is now tested as before, and the
process continued until an inverse is found such that each element of
array C is less than a given small quantity.

9. Write a procedure to find the inverse of a matrix using a method of
maximum pivots. Use the array A[1:n, 1:2×n] to store both the
matrix and its inverse. A supplementary array will be needed to record
interchanges of columns and to unscramble the inverse before it is printed
(see program 10.6).

MISCELLANEOUS PROBLEMS

1. (a) Simple interest is paid on £100 invested at r% for n years. Write a program to print an interest table for values of r from $\frac{1}{2}$% to $4\frac{1}{2}$% and for integral values of n between 1 and 20.

(b) Repeat question 1(a) for compound interest.

2. A mortgage of £1,000 is to be repaid in equal monthly instalments over a given period of years. The interest is calculated as a percentage of the loan outstanding at the time of each repayment. Write a program to print a table of monthly repayment charges for periods ranging from 1 to 25 years at rates of interest from 3% to 10%. (Consider only integral values of the variables.) It can be shown that the monthly repayment on a loan of £P over a period of n years is given by

$$\frac{P\, y^{12n}\,(1\, -\, y)}{1\, -\, y^{12n}}$$

where y is the growth rate per month. For example, if the interest is 5% p.a.

$$y = (1 + \tfrac{5}{1200})$$

3. (a) Draw up a table to show the amount of income tax paid by single men with salaries stepping by £100 from £1,000 to £3,000 p.a., assuming that they have no other source of income and claim no allowances. Draw up further tables for married men with 1, 2, 3 and 4 children, drawing maximum children's allowances. (See Tax Form No. P.3 or S.M.P. Book T.4, page 276.)

(b) Repeat question 3(a) but make allowance for National Insurance payments, superannuation at 6% of income p.a. and life insurance premiums of £40 p.a.

4. The co-ordinates of the vertices of two triangles are supplied as data. Write a program to find whether one triangle lies completely inside the boundary of the other.

5. Write a program to sort Roman numbers into descending order. The data tape is prepared using the input code given below, but output is to be in Roman numerals.

192

Integers greater than 1500 need not be considered.

Decimal number	1000	500	100	50	10	5	1
Roman equivalent	M	D	C	L	X	V	I
Input code	7	6	5	4	3	2	1

Example

The decimal numbers	609	748	and 999
would be output as	DCIX	DCCXLVIII	and CMXCIX
but punched as	6513;	655342111;	573513;

6. A factory is to be re-equipped, and a combination of machines is to be chosen from four different types, a, b, c and d. Each machine makes the same product and the maximum number of machines available for purchase is given below. The relative costs and operating data are also supplied, together with the constraints under which the factory must operate. Write a program to find all the combinations of available machines that give a profit of at least £5,000 p.a. and also the particular combination that provides the maximum profit.

Costs and operating data (per machine)	Machines				Constraints (maximum permitted total in each category)
	a	b	c	d	
Cost of installation in £	360	420	250	490	8000
Man-hours required per week	41	59	25	102	1400
Number of products per week	305	215	95	690	7000
Cost of operating in £ per week	22	43	19	37	800
Profit in £ per year	320	350	240	400	—
Number of available machines	4	9	6	8	20

7. Write a program to print out a calendar in a familiar form when supplied with an integer representing the year (e.g. 1973) on data tape. To calculate the day of a week, Zeller's congruence should be used:

$$F = \{[2.6m - 0.2] + K + D + [D/4] + [C/4] - 2C\} \mod 7$$

where square brackets represent the 'square bracket function' or whole number part, and mod 7 implies the remainder after division by 7.

193

F = day of week (Sunday = 0, Monday = 1, . . .)
K = day of month (1, 2, 3, . . . , 31)
C = century (18, 19, 20, . . .)
D = year in century (1, 2, 3, . . . , 99)
M = month number with January and February taken as months 11 and 12 of the preceding year (March = 1, April = 2, . . .)

Thus Feb. 8, 1967, has K = 8, C = 19 (not 20), D = 66, M = 12.

N.B. This formula only applies to years following the calendar change in 1752.

8. Players in a cricket team have batting figures for a season set out in a table as shown below:

Player's Number	Runs	Number of Innings	Times Not Out	Average
1	524	17	0	
2	677	14	2	
.	.	.	.	
.	.	.	.	
.	.	.	.	
15	39	7	4	
16	32	3	3	

Write a program to read figures set out in this form, to calculate the batting averages and to print the complete table in descending order of averages.

9. (a) If i is the lowest of four consecutive integers, each less than 100, write a program to find a value of i such that the sum of the cubes of the integers is itself a perfect cube.

(b) Write a program to investigate whether the cube of a given integer less than 20 can be expressed as the sum of the cubes of four other non-negative integers.

10. Write a procedure to find the area of a triangle given the co-ordinates of the three vertices.

11. A 'perfect number' is a number equal to the sum of all its divisors. For example, 6 and 28 are perfect numbers, because

$$6 = 1 + 2 + 3 \quad \text{and} \quad 28 = 1 + 2 + 4 + 7 + 14$$

Write a program to find all the perfect numbers less than 1000.

12. Consider sequences of the form $\dfrac{p_r}{q_r}, \dfrac{p_{r+1}}{q_{r+1}}, \dfrac{p_{r+2}}{q_{r+2}}$

where   (i) $p_{r+1} = p_r + 2q_r$ and $q_{r+1} = p_r + q_r$
and   (ii) $p_{r+1} = p_r^2 + q_r^2$ and $q_{r+1} = 2p_r q_r$

Take any positive integral starting values for p and q, print their successive values and show that in both sequences, p_r/q_r tend to the same value.

13. A continuous single-valued function has two consecutive zeros at a and b. Write procedures to find the length of arc between a and b and the volume generated when this arc is rotated through 360° about the x-axis.

14. Write a procedure ARC COS to calculate the principal value of the arc cos of an angle given in radians, using the Algol standard function arc tan (see program 9.2—Cubic).

15. Curve fitting. It is often useful to find the equation of a curve that best fits a given set of points. One way of finding such a curve is as follows. In the figure, P and Q are two of a given set of points and y = f(x) is a possible curve of best fit. P′ and Q′ are points on the curve having the same x-co-ordinates as the points P and Q. f(x) is chosen so that the sum of the squares of all the lengths PP′ and QQ′ is a minimum.

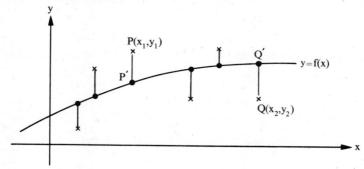

(a) **Linear case.** The equations for fitting a straight line y = ax + b to a given set of points (X, Y) are found to be:

$$a\Sigma X^2 + b\Sigma X = \Sigma XY$$

and
$$a\Sigma X + bN = \Sigma Y$$

where N is the number of points.

(b) **Quadratic case.** The corresponding equations for fitting a quadratic curve y = ax² + bx + c to a set of points (X, Y) are:

$$a\Sigma X^4 + b\Sigma X^3 + c\Sigma X^2 = \Sigma X^2 Y$$
$$a\Sigma X^3 + b\Sigma X^2 + c\Sigma X = \Sigma XY$$
$$a\Sigma X^2 + b\Sigma X + cN = \Sigma Y$$

Write programs to carry out curve fitting for a given set of points in each of the above cases.

195

16. (a) Write a program to output the digits of e to 100 places of decimals, using the following method, and truncating the series after the term $\frac{1}{70!}$.

Consider the calculation for e from the approximation

$$e = 1 + \frac{1}{1!} + \frac{1}{2!} + \frac{1}{3!} + \frac{1}{4!} + \frac{1}{5!} + \frac{1}{6!}$$

or

$$e = 2 + R_1$$

where

$$R_1 = \frac{1}{2!} + \frac{1}{3!} + \frac{1}{4!} + \frac{1}{5!} + \frac{1}{6!}.$$

It can be shown* that $R_1 < 1$, and so the decimal part of e is found by calculating R_1 only.

The first decimal place is found by multiplying the expression for R_1 by 10:

$$10R_1 = \frac{10}{2!} + \frac{10}{3!} + \frac{10}{4!} + \frac{10}{5!} + \frac{10}{6!}.$$

The numerators are now reduced, working from the right, in the following manner:

$$\frac{10}{6!} = \frac{1 \times 6 + 4}{6!} = \frac{1}{5!} + \frac{4}{6!}$$

The fraction $\frac{1}{5!}$ is now added to the term on its left to give $\frac{11}{5!}$, and this is expressed as

$$\frac{2 \times 5 + 1}{5!} = \frac{2}{4!} + \frac{1}{5!}$$

The process is continued until

$$10R_1 = 7 + \frac{0}{2!} + \frac{1}{3!} + \frac{0}{4!} + \frac{1}{5!} + \frac{4}{6!}$$

$$= 7 + R_2$$

Here again, $R_2 < 1$, and so the remainder, R_2, does not contribute to the value of the first decimal place, which is seen to be 7.

i.e.
$$e = 2 + \frac{7}{10} + \frac{R_2}{10}.$$

* The proof that the remainders R_1 at each stage are less than unity is as follows. The maximum possible remainder after reduction is

$$\frac{1}{2!} + \frac{2}{3!} + \frac{3}{4!} + \frac{4}{5!} + \ldots + \frac{n-1}{n!}$$

$$= \left(1 - \frac{1}{2!}\right) + \left(\frac{1}{2!} - \frac{1}{3!}\right) + \left(\frac{1}{3!} - \frac{1}{4!}\right) + \ldots + \left(\frac{1}{(n-1)!} - \frac{1}{n!}\right)$$

$$= 1 - \frac{1}{n!} \text{ which is less than 1.}$$

The expression for R_2 is now multiplied by 10 to give

$$10R_2 = \frac{0}{2!} + \frac{10}{3!} + \frac{0}{4!} + \frac{10}{5!} + \frac{40}{6!}$$

which is then reduced to

$$10R_2 = 1 + \frac{1}{2!} + \frac{2}{3!} + \frac{3}{4!} + \frac{1}{5!} + \frac{4}{6!}$$
$$= 1 + R_3 \text{ where } R_3 < 1$$

i.e.

$$e = 2 + \frac{7}{10} + \frac{1}{100} + \frac{R_3}{100}.$$

(b) Adapt the above program to output the digits of e five at a time by multiplying each numerator by 10^5 instead of 10.

17. Write a program to maximize a linear expression with m non-negative variables in the form $k_1x_1 + k_2x_2 + \ldots + k_mx_m$, given n constraints each in the form

$$c_1x_1 + c_2x_2 + \ldots + c_mx_m \leqslant c_0 \quad (c_0 > 0).$$

The method may be illustrated by the following example. Maximize $3x_1 + 2x_2$ given that $5x_1 + 3x_2 \leqslant 75$, $2x_1 + x_2 \leqslant 30$ and $x_1 + x_2 \leqslant 20$. The inequalities are first converted into equations by the addition of non-negative slack variables, x_3, x_4 and x_5:

$$5x_1 + 3x_2 + x_3 = 75 \tag{1}$$
$$2x_1 + x_2 + x_4 = 30 \tag{2}$$
$$x_1 + x_2 + x_5 = 20 \tag{3}$$

The equations are then rewritten with the slack variables as their subjects:

$$x_3 = 75 - 5x_1 - 3x_2 \tag{1}$$
$$x_4 = 30 - 2x_1 - x_2 \tag{2}$$
$$x_5 = 20 - x_1 - x_2 \tag{3}$$

The coefficients of the right-hand sides are written in the form of a matrix. A tableau is constructed by adding a top border containing the variables omitted from the matrix and a side border containing the slack variables.

A bottom border is added containing the coefficients of the expression to be maximized, together with a constant term of zero, (as shown in Tableau 1).

Each stage in the maximizing process entails a rearrangement of the original equations. The aim is to reduce the coefficients in the lower border (excluding the constant term, C) to negative or zero values. The stages in this reduction are represented by a series of tableaux, and the conversion from one tableau to the next is given by the following set of operations.

(a) Note the column (q) of the largest positive element in the lower border (excluding the constant term, C).

(b) For each row (i) of the matrix, find the ratio of the constant term to the term in column q. Note the row (p) of the negative ratio nearest to zero.

197

Tableau 1

	C	x_1	x_2
x_3	75	-5^*	-3
x_4	30	-2	-1
x_5	20	-1	-1
Max	0	3	2

The element A_{pq} is called the pivot (marked with an asterisk in the diagrams). In the event of more than one possible pivot, an arbitrary choice is made.

(c) Form a new tableau as follows:
  (1) replace each element A_{ij} ($i \neq p$, $j \neq q$) by $A_{ij} - A_{iq} \times A_{pj}/A_{pq}$,
  (2) in row p, replace each element A_{pj} ($j \neq q$) by $-A_{pj}/A_{pq}$,
  (3) in column q, replace each element A_{iq} ($i \neq p$) by A_{iq}/A_{pq},
  (4) replace pivot A_{pq} by $1/A_{pq}$.

(d) Interchange the variable in column q of the top border with the variable in row p of the left border.

In the above example, two rearrangements are required, and the results of the operations are shown below.

Tableau 2

	C	x_3	x_2
x_1	15	$-\frac{1}{5}$	$-\frac{3}{5}$
x_4	0	$\frac{2}{5}$	$\frac{1}{5}$
x_5	5	$\frac{1}{5}$	$-\frac{2}{5}^*$
Max	45	$-\frac{3}{5}$	$\frac{1}{5}$

Tableau 3

	C	x_3	x_5
x_1	$7\frac{1}{2}$	$-\frac{1}{2}$	$\frac{3}{2}$
x_4	$2\frac{1}{2}$	$\frac{1}{2}$	$-\frac{1}{2}$
x_2	$12\frac{1}{2}$	$\frac{1}{2}$	$-\frac{5}{2}$
Max	$47\frac{1}{2}$	$-\frac{1}{2}$	$-\frac{1}{2}$

The process is terminated at Tableau 3 where all the coefficients other than the constant term in the bottom border are negative or zero. The values of x_1, x_4 and x_2 are read from column C of the final tableau, and the maximum value of $3x_1 + 2x_2$ is found to be $47\frac{1}{2}$. In each tableau, the variables not appearing in the left border are zero.

18. The 'Tower of Hanoi' problem requires n discs, each with a different diameter. The discs may be piled in any of three locations, but at no time may a disc be placed on top of a smaller disc. If the discs are initially piled

in the correct order in location number 1, write a recursive procedure to
record the moves necessary to transfer the discs, one at a time, until they are
piled in the correct order in location number 2. For example, if n = 3,
the required moves are

$$1 \to 2$$
$$1 \to 3$$
$$2 \to 3$$
$$1 \to 2$$
$$3 \to 1$$
$$3 \to 2$$
$$1 \to 2$$

The instruction $2 \to 3$ implies a move of the top disc from the pile in location 2 to that in location 3.

SOLUTIONS TO EXERCISES (CHAPTER 2)

1. (a) $+22.3$ 22.3 $1_{10}-7$ $_{10}-7$ $10_{10}7$ $-3_{10}2$
   (b) six   true   Highest factor   XVII   counter   goto   delta   Open Lambda   sine   P8C
   (c) W↑D   a + (b)   product/two
   (d) sqrt (abs(—3.5))   ln (0.2)

2. $3 + 4$ $7 \div 4$ $4↑3$ $12 \div 4↑3$ sign (42/13)   entier (4.3) sign (sqrt (7))

3. (a) 3   (b) 0   (c) 3   (d) 3   (e) 10   (f) $1\frac{1}{2}$ (g) $7\frac{1}{2}$ (h) -2 (i) 2 (j) -2.

4. Parts (a), (b), (c), (d).

(e)

x	0	7	12	15	16
i	(9)7	5	3	1	.

i	.	-2	-2	-2	-2	1	1	1	1	4	4	4	4
j	.	3	1	-1	-3	3	1	-1	-3	3	1	-1	-3
sum	0	-6	-8	-6	0	3	4	3	0	12	16	12	0

SUGGESTIONS FOR FURTHER READING

Battersby, A: *Mathematics in Management* (Pelican).

Burnett-Hall, D.G., Dresel, L.A.G. and Samet, P.A.: *Computer Programming and Autocodes* (E.U.P.).

Gruenberger, F. and Jaffray, G.: *Problems for Computer Solution* (Wiley).

Hawgood, J.: *Numerical Methods in Algol* (McGraw-Hill).

Marchant, J. P. and Pegg, D.: *Digital Computers* (Blackie).

Moakes, A.J.: *Numerical Mathematics* (Macmillan).

Noble, B.: *Numerical Methods; I and II* (Oliver and Boyd).

Pennington, R.H.: *Introductory Computer Methods and Numerical Analysis* (Macmillan).

Redish, K.A.: *Computational Methods* (E.U.P.)

INDEX